The Volume 2 of Feng(Funn)'s Mathematics

Xu Feng (Funn Heo)

Yes, it's time to write something for the Volume 2 of Feng(Funn)'s Mathematics. You all know that the Volume 2 is a gift to Dinbaken for lots of geniuses from Dinbak.

Now, you also know that some words in Romanization are: Xu Feng = Funn Heo, which means my name. Dianbai = Dinbak, which means my country and my nation. Dianbaian = Dinbakan (singular n), which means one of Dinbak's people. Dianbaien = Dinbaken (Plural n), which mean all of Dinbak's people. Dianbaish = Dinbakish, which is adjective. A Dinbakan = A Dinbakish person; Dinbaken = the Dinbakish people. So, I like you call me Funn Heo.

The other words in Romanization are: In Cantonish, Guangdong = Cantonia, Guangdongen = Cantonians. And Cantonich is adjective. A Cantonich Person = A Cantonian. Cantonian that means one of Cantonia's people. Cantonians that mean all of Cantonia's people. Cantonians = the Cantonich people. So Cantonians are, too, not Chinese. Sometimes I think Cantonians still don't know the history of themselves. Barracoon that is a witnessing word is shame on Cantonians. While Cantonians want to be Chinese, they are always slaves. That's all. In other words, Cantonians need to grow up. They are just the Cantonich people, not the other people. They have to go fighting for themselves, and this is one of their responsibilities.

Sure, the fucking Chinese language is shit and inferiority. It just suits the fucking China and the fucking Chinese.

The other wonderful thing is that the Chinese are hopeless anytime and anywhere. it is crazy, right? Good. Just for this fact, we have to cheers. Haha…

Well, Dinbaken, why do we use the Latin letters to write everything? One of the reasons is these:
1) When you know that: C, C++, Pascal and the other languages of the computers are not better than you do, the Latin letters help you to make the new languages of the computers.
2) Also, when you know that: Windows, Android, IOS and the other operating systems are not better than you do, the Latin letters help you to make the new operating systems.

Today, let us look back at the past human history, we Dinbaken have no rights to grumble. We know that we Dinbaken owe this world lots of geniuses, not the others.

Sometimes, Dinbaken, when you look at the Periodic Table, what do you think? The problem is that: Without your names in the Periodic Table, it's not perfect and great. So, ones of you Dinbaken, can you put your names in the Periodic Table? Of course, I know the people envy you, and you make me proud.

Who can write my biography? You Dinbaken. But it has a qualification: When Dinbak has enough geniuses born in, now you can start to write my biography in English or in the other languages, not in the fucking Chinese. You know that I dislike the fucking Chinese and its fucking language.

We Dinbaken are not Asians. We Dinbaken are just the Dinbakish people. We Dinbaken don't care who are Asians. Also, we Dinbaken don't care who are Europeans. All I can thank Europe is Latin letters and its grammars.

The Anglo-Saxons are also losers in the world. The USA is dying now, and the UK faces the same trouble just like the USA. The other crazy thing is that: do you know why Canada and Australia lose their vitalities in their future? Because of the Chinese, that's all.

And what did I say, Europeans? Two choices in front of you Europeans: ① Five new countries in America; ② Six new countries in America. You Europeans make a decision: ① You get an African-

European country; ② You get the European country. Especially you are Anglo-Saxons, just thinking about England. Don't waste your time.

Well, maybe Europeans want to live in the Europeantown, that should be super great. Haha...but you Europeans know that: Cantonians are not the fucking Chinese. Why do you Europeans let Cantonians live in the fucking Chinatown?

One bad thing that is colonialism; the other bad thing that is immigration policy.

And now, Cantonians, one thing you must know that: Dinbak is near Cantonia, not in Cantonia. So, Cantonians and Dinbaken are different. How many times I asked myself: Cantonians and Dinbaken might be together in a nation and a country, but in fact, it's wrong. You Cantonians go your way, we Dinbaken go our way.

All human beings, you know that Dinbak is my nation and my country. It's just near Asia, and never belongs to anywhere. If you don't know what is Dinbakish Culture, it means that you are no human beings, you are shits. In front of us Dinbaken, don't talk about how great you are or you were. You know why? You all like the fucking Chinese cry in the world so poor. Ha. And, Dinbaken, I make you proud anywhere and anytime. Also, of course, you Dinbaken make me proud anywhere and anytime. Just let us cheer for this forever. Haha…

I can give Cantonians a gift that is on Jyutping. It means Cantonish in Romanization, but I'm not sure what you Cantonians think of your language. I wrote down on Jyutping because the other languages in my heart are not better than I do. Another reason is for my people Dinbaken. So, Cantonish like English shall be a second language of Dinbaken, and always be.

A question: Cantonians, do you want to make yourselves great? Using the Latin Letters to write everything in your lives, this is only one way you go. The other question: Cantonians, do you allow that the fucking Chinese go on to colonize Cantonia? Would you like to be governing by the fucking Chinese? But don't tell me that you would love to be slaves, okay? That is a morbid state.

Adding some words to Russians: Mongol Empire was gone a long time, why do you still let your hearts stay in barbarism? You also need to grow up: Be kindness with civilization.

In this Volume, it has three parts. Part 1: ① Dinbakish Equations; ② Composite Dinbakish Equations; ③ Multivariate Dinbakish Equations; ④ Composite-Multivariate Dinbakish Equations. Part 2: ① Dinbakish Theorems; ② Multivariate Funn Equations; ③ Composite-Multivariate Dinbakish-Funn Equations. Part 3: Dinbakish Equations and their Inverse Functions.

P.S.   A long time (8 years) I didn't use MS Office if OpenOffice was working fine. But today just look at that LibreOffice is not working fine anymore on my computer, it is a helldamn. Ha.

Part 1

1:

Looking at $\left(x^{\frac{1}{2}}+2\right)^4 = 0$, it expands: $1x^2 + 8x^{\frac{3}{2}} + 20x + 32x^{\frac{1}{2}} + 16 = 0$;

Now, let the coefficient number 1 to be a, the coefficient number 8 to be b, the coefficient number 20 to be c, the coefficient number 32 to be d, and the integer number 16 to be e, so that the equation

$1x^2 + 8x^{\frac{3}{2}} + 20x + 32x^{\frac{1}{2}} + 16 = 0$ becomes $ax^2 + bx^{\frac{3}{2}} + cx + dx^{\frac{1}{2}} + e = 0$.

$\left(a \in R, b \in R, c \in R, d \in R, e \in R, a \neq 0, b \neq 0, c \neq 0, d \neq 0, e \neq 0\right)$.

2:

Also looking at $\left(x^{\frac{1}{2}}+2\right)^6 = 0$, it expands:

$C_6^0 x^3 + 2C_6^1 x^{\frac{5}{2}} + 4C_6^2 x^2 + 8C_6^3 x^{\frac{3}{2}} + 16C_6^4 x + 32C_6^5 x^{\frac{1}{2}} + 64C_6^6 = 0$;

Now, let the coefficient number C_6^0 to be a, the coefficient number $2C_6^1$ to be b, the coefficient number $4C_6^2$ to be c, the coefficient number $8C_6^3$ to be d, the coefficient number $16C_6^4$ to be e, the coefficient number $32C_6^5$ to be f, and the integer number $64C_6^6$ to be g.

The equation $C_6^0 x^3 + 2C_6^1 x^{\frac{5}{2}} + 4C_6^2 x^2 + 8C_6^3 x^{\frac{3}{2}} + 16C_6^4 x + 32C_6^5 x^{\frac{1}{2}} + 64C_6^6 = 0$

becomes $ax^3 + bx^{\frac{5}{2}} + cx^2 + dx^{\frac{3}{2}} + ex + fx^{\frac{1}{2}} + g = 0$.

$\left(a \in R, b \in R, c \in R, d \in R, e \in R, f \in R, g \in R, a \neq 0, b \neq 0, c \neq 0, d \neq 0, e \neq 0, f \neq 0, g \neq 0\right)$

3: Making a summary from 1 to 2, they have a regularity:

$$a_n x^{(n)\frac{1}{2}} + a_{n-1} x^{(n-1)\frac{1}{2}} + a_{n-2} x^{(n-2)\frac{1}{2}} + \ldots + a_2 x + a_1 x^{\frac{1}{2}} + a_0 = 0 \,,\ (n = 2,3,4,5,\ldots,\infty)\,.$$

In an equation, when its index numbers are the fractions, which are an arithmetic progression, and its common difference is a fraction, I call it the Dinbakish Equation.

(1): According to the Dinbakish Equation, lots of different forms in which

they are:

I : $\ a_n x^{(n)\frac{1}{k}} + a_{n-1} x^{(n-1)\frac{1}{k}} + a_{n-2} x^{(n-2)\frac{1}{k}} + \ldots + a_1 x^{\frac{1}{k}} + a_0 = 0 \,;$

II : $\ a_n x^{(n)\frac{j}{k}} + a_{n-1} x^{(n-1)\frac{j}{k}} + a_{n-2} x^{(n-2)\frac{j}{k}} + \ldots + a_1 x^{\frac{j}{k}} + a_0 = 0 \,;$

III : $\ a_n x^{(n)(-\frac{1}{k})} + a_{n-1} x^{(n-1)(-\frac{1}{k})} + a_{n-2} x^{(n-2)(-\frac{1}{k})} + \ldots + a_1 x^{-\frac{1}{k}} + a_0 = 0 \,;$

IV: $\ a_n x^{(n)(-\frac{j}{k})} + a_{n-1} x^{(n-1)(-\frac{j}{k})} + a_{n-2} x^{(n-2)(-\frac{j}{k})} + \ldots + a_1 x^{-\frac{j}{k}} + a_0 = 0 \,.$

In which $n = 1,2,3,4,\ldots,\infty$; $k = 2,3,4,5,\ldots,\infty$; $j = 2,3,4,5,\ldots,\infty$.

For deepening to understand the Dinbakish Equations:

①: A Dinbakish Equation

$$ax^3 + bx^{\frac{5}{2}} + cx^2 + dx^{\frac{3}{2}} + ex + fx^{\frac{1}{2}} + g = 0 \,,$$

$(a \in R, b \in R, c \in R, d \in R, e \in R, f \in R, g \in R, a \neq 0, b \neq 0, c \neq 0, d \neq 0, e \neq 0, f \neq 0, g \neq 0)$

4

(a): One of the forms of the Dinbakish Equation can be:

$$a\left(x^{\frac{1}{2}}-2a\right)\left(x^{\frac{1}{2}}+\frac{a}{2}\right)\left(x^{\frac{1}{2}}+\frac{1}{a}\right)\left(x^{\frac{1}{2}}+\frac{1}{2a}\right)\left(x^{\frac{1}{2}}-2\right)\left(x^{\frac{1}{2}}-3\right)=0,$$

It expands:

$$ax^3+\left(\frac{3}{2}-\frac{3a^2}{2}-5a\right)x^{\frac{5}{2}}+\left[6a-5\left(\frac{3}{2}-\frac{3a^2}{2}\right)+\frac{1}{2a}-\frac{9a}{4}-a^3\right]x^2+\left[6\left(\frac{3}{2}-\frac{3a^2}{2}\right)-5\left(\frac{1}{2a}-\frac{9a}{4}-a^3\right)-\left(\frac{3}{4}+\frac{3a^2}{2}\right)\right]x^{\frac{3}{2}}+\left[6\left(\frac{1}{2a}-\frac{9a}{4}-a^3\right)-\frac{a}{2}+5\left(\frac{3}{4}+\frac{3a^2}{2}\right)\right]x+\left[\frac{5a}{2}-6\left(\frac{3}{4}+\frac{3a^2}{2}\right)\right]x^{\frac{1}{2}}-3a=0,$$

And $b=\frac{3}{2}-\frac{3a^2}{2}-5a$, $c=6a-5\left(\frac{3}{2}-\frac{3a^2}{2}\right)+\frac{1}{2a}-\frac{9a}{4}-a^3$,

$$d=6\left(\frac{3}{2}-\frac{3a^2}{2}\right)-5\left(\frac{1}{2a}-\frac{9a}{4}-a^3\right)-\left(\frac{3}{4}+\frac{3a^2}{2}\right), \quad e=6\left(\frac{1}{2a}-\frac{9a}{4}-a^3\right)-\frac{a}{2}+5\left(\frac{3}{4}+\frac{3a^2}{2}\right),$$

$$f=\frac{5a}{2}-6\left(\frac{3}{4}+\frac{3a^2}{2}\right), \quad g=-3a.$$

The Dinbakish Equation follows the real number a to be changed.

When $a=1$, it means that: $b=-5$, $c=\frac{13}{4}$, $d=\frac{23}{2}$, $e=-\frac{23}{4}$,

$$f=-11, \quad g=-3.$$

Now, the Dinbakish Equation is:

$$x^3-5x^{\frac{5}{2}}+\frac{13}{4}x^2+\frac{23}{2}x^{\frac{3}{2}}-\frac{23}{4}x-11x^{\frac{1}{2}}-3=0,$$

And its roots are: 1) $x^{\frac{1}{2}}-2=0$, $x_1=4$; 2) $x^{\frac{1}{2}}+\frac{1}{2}=0$, $x_2=\left(-\frac{1}{2}\right)^2$;

3) $x^{\frac{1}{2}}+1=0$; $x_3=(-1)^2$; 4) $x^{\frac{1}{2}}-2=0$, $x_4=4$; 5) $x^{\frac{1}{2}}-3=0$, $x_5=9$;

6) $x^{\frac{1}{2}}+\frac{1}{2}=0$, $x_6=\left(-\frac{1}{2}\right)^2$.

(b): One of the forms of the Dinbakish Equation can be:

$$\left(ax^{\frac{1}{2}}-5\right)\left(x^{\frac{1}{2}}-a\right)\left(x^{\frac{1}{2}}+a\right)\left(x^{\frac{1}{2}}+2a\right)\left(x^{\frac{1}{2}}-2a\right)\left(x^{\frac{1}{2}}+3a\right)=0;$$

It expands:

$$ax^3+\left(3a^2-5\right)x^{\frac{5}{2}}-\left(15a+5a^3\right)x^2-5\left(3a^2-5\right)x^{\frac{3}{2}}+\left(75a^3+4a^4\right)x+4a^4\left(3a^2-5\right)x^{\frac{1}{2}}-60a^5=0,$$

and $b=3a^2-5$, $c=-\left(15a+5a^3\right)$, $d=-5\left(3a^2-5\right)$, $e=75a^3+4a^4$,

$$f=4a^4\left(3a^2-5\right), \ g=-60a^5.$$

The Dinbakish Equation follows the real number a to be changed.

When $a=-1$, it means that:

$b=-2$, $c=20$, $d=10$, $e=-71$, $f=-8$, $g=60$.

The Dinbakish Equation is:

$$-x^3-2x^{\frac{5}{2}}+20x^2+10x^{\frac{3}{2}}-71x-8x^{\frac{1}{2}}+60=0,$$

Its roots are: 1) $x^{\frac{1}{2}}+5=0$, $x_1=(-5)^2$; 2) $x^{\frac{1}{2}}+1=0$, $x_2=(-1)^2$;

3) $x^{\frac{1}{2}}-1=0$, $x_3=1$; 4) $x^{\frac{1}{2}}+2=0$, $x_4=(-2)^2$;

6

5) $x^{\frac{1}{2}} - 2 = 0$, $x_5 = 4$; 6) $x^{\frac{1}{2}} - 3 = 0$, $x_6 = 9$.

(c): Why does the Dinbakish Equation have the infinite forms of itself?

This is one of the reasons:

$$\left(ax^{\frac{1}{2}} \pm k\right)\left(x^{\frac{1}{2}} \pm ka\right)\left(x^{\frac{1}{2}} \pm 2ka\right)\left(x^{\frac{1}{2}} \pm a\right)\left(x^{\frac{1}{2}} \pm 3ka\right)\left(x^{\frac{1}{2}} \pm 4ka\right)\left(x^{\frac{1}{2}} \pm 5ka\right) = 0,$$

$$\left(k \in R, a \in R, k \neq 0, a \neq 0\right).$$

②: A Dinbakish Equation

$$ax^{10} + bx^{\frac{15}{2}} + cx^5 + dx^{\frac{5}{2}} + e = 0,$$

$$\left(a \in R, b \in R, c \in R, d \in R, e \in R, a \neq 0, b \neq 0, c \neq 0, d \neq 0, e \neq 0\right)$$

One of the forms of the Dinbakish Equation can be:

$$a\left(x^{\frac{5}{2}} - 3a\right)\left(x^{\frac{5}{2}} - 4a\right)\left(x^{\frac{5}{2}} - \frac{a}{2}\right)\left(x^{\frac{5}{2}} + \frac{2}{3a}\right) = 0,$$

It expands:

$$ax^{10} + \left(\frac{2}{3} - \frac{a^2}{2} - 7a^2\right)x^{\frac{15}{2}} + \left[12a^3 - \frac{a}{3} - 7a^2\left(\frac{2}{3a} - \frac{a}{2}\right)\right]x^5 + \left[12a^3\left(\frac{2}{3a} - \frac{a}{2}\right) + \frac{7a^2}{3}\right]x^{\frac{5}{2}} - 4a^3 = 0,$$

and $b = \frac{2}{3} - \frac{a^2}{2} - 7a^2$, $c = 12a^3 - \frac{a}{3} - 7a^2\left(\frac{2}{3a} - \frac{a}{2}\right)$,

$$d = 12a^3\left(\frac{2}{3a} - \frac{a}{2}\right) + \frac{7a^2}{3}, \quad e = -4a^3.$$

The Dinbakish Equation follows the real number a to be changed.

When $a = \sqrt{2}$, it means that:

$$b = -\frac{43}{3}, \quad c = 26\sqrt{2}, \quad d = -\frac{10}{3}, \quad e = -8\sqrt{2}.$$

Now, the Dinbakish Equation is:

$$\sqrt{2}x^{10} - \frac{43}{3}x^{\frac{15}{2}} + 26\sqrt{2}x^5 - \frac{10}{3}x^{\frac{5}{2}} - 8\sqrt{2} = 0,$$

Its roots are: 1) $x^{\frac{5}{2}} - 3\sqrt{2} = 0$, $x_1 = \left(3\sqrt{2}\right)^{\frac{2}{5}}$; 2) $x^{\frac{5}{2}} - 4\sqrt{2} = 0$, $x_2 = \left(4\sqrt{2}\right)^{\frac{2}{5}}$;

3) $x^{\frac{5}{2}} - \frac{\sqrt{2}}{2} = 0$, $x_3 = \left(\frac{\sqrt{2}}{2}\right)^{\frac{2}{5}}$; 4) $x^{\frac{5}{2}} + \frac{\sqrt{2}}{3} = 0$, $x_4 = \left(-\frac{\sqrt{2}}{3}\right)^{\frac{2}{5}}$.

③: A Dinbakish Equation

$$ax^{-\frac{5}{3}} + bx^{-\frac{4}{3}} + cx^{-1} + dx^{-\frac{2}{3}} + ex^{-\frac{1}{3}} + f = 0,$$

$$\left(a \in R, b \in R, c \in R, d \in R, e \in R, f \in R, a \neq 0, b \neq 0, c \neq 0, d \neq 0, e \neq 0, f \neq 0\right)$$

One of the forms of the Dinbakish Equation can be:

$$a\left(x^{-\frac{1}{3}} + 3a\right)\left(x^{-\frac{1}{3}} - 4a\right)\left(x^{-\frac{1}{3}} - \frac{a}{2}\right)\left(x^{-\frac{1}{3}} + \frac{3}{a}\right)\left(x^{-\frac{1}{3}} + \frac{3a}{2}\right) = 0,$$

It expands:

$$ax^{-\frac{5}{3}} + 3x^{-\frac{4}{3}} - \frac{55a^3}{4}x^{-1} - \frac{165a^2 + 45a^4}{4}x^{-\frac{2}{3}} + \left(9a^5 - \frac{135a^3}{4}\right)x^{-\frac{1}{3}} + 27a^4 = 0,$$

and $b=3$, $c=-\dfrac{55a^3}{4}$, $d=-\dfrac{165a^2+45a^4}{4}$, $e=9a^5-\dfrac{135a^3}{4}$, $f=27a^4$.

The Dinbakish Equation follows the real number a to be changed.

When $a=-3$, it means that:

$$c=\dfrac{1485}{4}, \quad d=-\dfrac{5130}{4}, \quad e=-\dfrac{5103}{4}, \quad f=2187.$$

Now, the Dinbakish Equation is:

$$-3x^{-\frac{5}{3}}+3x^{-\frac{4}{3}}+\dfrac{1485}{4}x^{-1}-\dfrac{5130}{4}x^{-\frac{2}{3}}-\dfrac{5103}{4}x^{-\frac{1}{3}}+2187=0,$$

Its roots are: 1) $x^{-\frac{1}{3}}-9=0$, $x_1=9^{-3}$; 2) $x^{-\frac{1}{3}}+12=0$, $x_2=(-12)^{-3}$;

3) $x^{-\frac{1}{3}}+\dfrac{3}{2}=0$, $x_3=\left(-\dfrac{3}{2}\right)^{-3}$; 4) $x^{-\frac{1}{3}}-1=0$, $x_4=1$; 5) $x^{-\frac{1}{3}}-\dfrac{9}{2}=0$, $x_5=\left(\dfrac{9}{2}\right)^{-3}$.

④: A Dinbakish Equation

$$ax^{-\frac{9}{2}}+bx^{-\frac{15}{4}}+cx^{-3}+dx^{-\frac{9}{4}}+ex^{-\frac{3}{2}}+fx^{-\frac{3}{4}}+g=0,$$

$\left(a\in R, b\in R, c\in R, d\in R, e\in R, f\in R, g\in R, a\neq0, b\neq0, c\neq0, d\neq0, e\neq0, f\neq0, g\neq0\right)$

One of the forms of the Dinbakish Equation can be:

$$a\left(x^{-\frac{3}{4}}-3a\right)\left(x^{-\frac{3}{4}}-\dfrac{2}{3a}\right)\left(x^{-\frac{3}{4}}-5a\right)\left(x^{-\frac{3}{4}}-6a\right)\left(x^{-\frac{3}{4}}-\dfrac{1}{a}\right)\left(x^{-\frac{3}{4}}-2a\right)=0,$$

It expands:

$$ax^{-\frac{9}{2}} - \left(16a^2 + \frac{5}{3}\right)x^{-\frac{15}{4}} + \left(91a^3 + \frac{80a}{3} + \frac{2}{3a}\right)x^{-3} - \left(216a^4 + \frac{455a^2}{3} + \frac{32}{3}\right)x^{-\frac{9}{4}} + \left(360a^3 + 180a^5 + \frac{182a}{3}\right)x^{-\frac{3}{2}} - \left(300a^4 + 144a^2\right)x^{-\frac{3}{4}} + 120a^3 = 0$$

and $b = -\left(16a^2 + \frac{5}{3}\right)$, $c = 91a^3 + \frac{80a}{3} + \frac{2}{3a}$, $d = -\left(216a^4 + \frac{455a^2}{3} + \frac{32}{3}\right)$,

$$e = 360a^3 + 180a^5 + \frac{182a}{3}, \quad f = -\left(300a^4 + 144a^2\right), \quad g = 120a^3.$$

The Dinbakish Equation follows the real number a to be changed.

When $a = -\frac{1}{2}$, it means that:

$$b = -\frac{17}{3}, \quad c = -\frac{625}{24}, \quad d = -\frac{745}{12}, \quad e = -\frac{1943}{24}, \quad f = -\frac{219}{4}, \quad g = -15.$$

Now, the Dinbakish Equation is:

$$-\frac{1}{2}x^{-\frac{9}{2}} - \frac{17}{3}x^{-\frac{15}{4}} - \frac{625}{24}x^{-3} - \frac{745}{12}x^{-\frac{9}{4}} - \frac{1943}{24}x^{-\frac{3}{2}} - \frac{219}{4}x^{-\frac{3}{4}} - 15 = 0,$$

Its roots are: 1) $x^{-\frac{3}{4}} + \frac{3}{2} = 0$, $x_1 = \left(-\frac{3}{2}\right)^{-\frac{4}{3}}$; 2) $x^{-\frac{3}{4}} + \frac{4}{3} = 0$, $x_2 = \left(-\frac{4}{3}\right)^{-\frac{4}{3}}$;

3) $x^{-\frac{3}{4}} + \frac{5}{2} = 0$, $x_3 = \left(-\frac{5}{2}\right)^{-\frac{4}{3}}$; 4) $x^{-\frac{3}{4}} + 3 = 0$, $x_4 = (-3)^{-\frac{4}{3}}$;

5) $x^{-\frac{3}{4}} + 2 = 0$, $x_5 = (-2)^{-\frac{4}{3}}$; 6) $x^{-\frac{3}{4}} + 1 = 0$, $x_6 = (-1)^{-\frac{4}{3}}$.

(2): The other Dinbakish Equations

I: $a_n x^{(n)\frac{\sqrt{j}}{k}} + a_{n-1} x^{(n-1)\frac{\sqrt{j}}{k}} + a_{n-2} x^{(n-2)\frac{\sqrt{j}}{k}} + \ldots + a_2 x^{(2)\frac{\sqrt{j}}{k}} + a_1 x^{\frac{\sqrt{j}}{k}} + a_0 = 0;$

$$\text{II:}\quad a_n x^{(n)(-\frac{\sqrt{j}}{k})} + a_{n-1} x^{(n-1)(-\frac{\sqrt{j}}{k})} + a_{n-2} x^{(n-2)(-\frac{\sqrt{j}}{k})} + \ldots + a_2 x^{(2)(-\frac{\sqrt{j}}{k})} + a_1 x^{-\frac{\sqrt{j}}{k}} + a_0 = 0 \, \dot{;}$$

$$\text{III:}\quad a_n x^{(n)\frac{i\sqrt{j}}{k}} + a_{n-1} x^{(n-1)\frac{i\sqrt{j}}{k}} + a_{n-2} x^{(n-2)\frac{i\sqrt{j}}{k}} + \ldots + a_2 x^{(2)\frac{i\sqrt{j}}{k}} + a_1 x^{\frac{i\sqrt{j}}{k}} + a_0 = 0 \, \dot{;}$$

$$\text{IV:}\quad a_n x^{(n)(-\frac{i\sqrt{j}}{k})} + a_{n-1} x^{(n-1)(-\frac{i\sqrt{j}}{k})} + a_{n-2} x^{(n-2)(-\frac{i\sqrt{j}}{k})} + \ldots + a_2 x^{(2)(-\frac{i\sqrt{j}}{k})} + a_1 x^{-\frac{i\sqrt{j}}{k}} + a_0 = 0 \, .$$

$$\left(i = 2,3,4,\ldots,\infty ; \, j = 2,3,4,\ldots,\infty ; \, k = 2,3,4,\ldots,\infty ; \, n = 2,3,4,\ldots,\infty \right)$$

①: A Dinbakish Equation

$$ax^{\frac{5\sqrt{2}}{4}} + bx^{\sqrt{2}} + cx^{\frac{3\sqrt{2}}{4}} + dx^{\frac{\sqrt{2}}{2}} + ex^{\frac{\sqrt{2}}{4}} + f = 0 \, ,$$

$$\left(a \in R, b \in R, c \in R, d \in R, e \in R, f \in R, a \neq 0, b \neq 0, c \neq 0, d \neq 0, e \neq 0, f \neq 0 \right)$$

One of the forms of the Dinbakish Equation can be:

$$\left(ax^{\frac{\sqrt{2}}{4}} - 1 \right)\left(x^{\frac{\sqrt{2}}{4}} - a \right)\left(x^{\frac{\sqrt{2}}{4}} + a \right)\left(x^{\frac{\sqrt{2}}{4}} - \frac{1}{2a} \right)\left(x^{\frac{\sqrt{2}}{4}} + 2a \right) = 0 \, ,$$

It expands:

$$ax^{\frac{5\sqrt{2}}{4}} + \left(2a^2 - \frac{3}{2} \right)x^{\sqrt{2}} - \left(a^3 + 3a - \frac{1}{2a} \right)x^{\frac{3\sqrt{2}}{4}} - \left(2a^4 - \frac{3a^2}{2} - 1 \right)x^{\frac{\sqrt{2}}{2}} + \left(3a^3 - \frac{a}{2} \right)x^{\frac{\sqrt{2}}{4}} - a^2 = 0 \, ,$$

11

and $b = 2a^2 - \dfrac{3}{2}$, $c = -\left(a^3 + 3a - \dfrac{1}{2a}\right)$, $d = -\left(2a^4 - \dfrac{3a^2}{2} - 1\right)$,

$e = 3a^3 - \dfrac{a}{2}$, $f = -a^2$.

The Dinbakish Equation follows the real number a to be changed.

When $a = 1$, it means that:

$b = \dfrac{1}{2}$, $c = -\dfrac{7}{2}$, $d = \dfrac{1}{2}$, $e = \dfrac{5}{2}$, $f = -1$.

The Dinbakish Equation is:

$$x^{\frac{5\sqrt{2}}{4}} + \frac{1}{2}x^{\sqrt{2}} - \frac{7}{2}x^{\frac{3\sqrt{2}}{4}} + \frac{1}{2}x^{\frac{\sqrt{2}}{2}} + \frac{5}{2}x^{\frac{\sqrt{2}}{4}} - 1 = 0,$$

Its roots are: 1) $x^{\frac{\sqrt{2}}{4}} - 1 = 0$, $x_1 = 1$; 2) $x^{\frac{\sqrt{2}}{4}} - 1 = 0$, $x_2 = 1$;

3) $x^{\frac{\sqrt{2}}{4}} + 1 = 0$, $x_3 = (-1)^{2\sqrt{2}}$; 4) $x^{\frac{\sqrt{2}}{4}} - \frac{1}{2} = 0$, $x_4 = \left(\frac{1}{2}\right)^{2\sqrt{2}}$,

5) $x^{\frac{\sqrt{2}}{4}} + 2 = 0$, $x_5 = (-2)^{2\sqrt{2}}$.

②: A Dinbakish Equation

$$ax^{-\sqrt{3}} + bx^{-\frac{4\sqrt{3}}{5}} + cx^{-\frac{3\sqrt{3}}{5}} + dx^{-\frac{2\sqrt{3}}{5}} + ex^{-\frac{\sqrt{3}}{5}} + f = 0,$$

$$\left(a \in R, b \in R, c \in R, d \in R, e \in R, f \in R, a \neq 0, b \neq 0, c \neq 0, d \neq 0, e \neq 0, f \neq 0\right)$$

One of the forms of the Dinbakish Equation can be:

$$a\left(x^{-\frac{\sqrt{3}}{5}} - \frac{1}{a}\right)\left(x^{-\frac{\sqrt{3}}{5}} + \frac{1}{a}\right)\left(x^{-\frac{\sqrt{3}}{5}} - \frac{2}{a}\right)\left(x^{-\frac{\sqrt{3}}{5}} + \frac{2}{a}\right)\left(x^{-\frac{\sqrt{3}}{5}} - 2a\right) = 0,$$

It expands:

$$ax^{-\sqrt{3}} - 2a^2x^{-\frac{4\sqrt{3}}{5}} - \frac{5}{a}x^{-\frac{3\sqrt{3}}{5}} + 10x^{-\frac{2\sqrt{3}}{5}} + \frac{4}{a^3}x^{-\frac{\sqrt{3}}{5}} - \frac{8}{a^2} = 0,$$

and $b = -2a^2$, $c = -\dfrac{5}{a}$, $d = 10$, $e = \dfrac{4}{a^3}$, $f = -\dfrac{8}{a^2}$.

The Dinbakish Equation follows the real number a to be changed.

When $a = 5$, it means that:

$b = -50$, $c = -1$, $e = \dfrac{4}{125}$, $f = -\dfrac{8}{25}$.

Now, the Dinbakish Equation is:

$$5x^{-\sqrt{3}} - 50x^{-\frac{4\sqrt{3}}{5}} - x^{-\frac{3\sqrt{3}}{5}} + 10x^{-\frac{2\sqrt{3}}{5}} + \frac{4}{125}x^{-\frac{\sqrt{3}}{5}} - \frac{8}{25} = 0,$$

Its roots are: 1) $x^{-\frac{\sqrt{3}}{5}} - \frac{1}{5} = 0$, $x_1 = \left(\frac{1}{5}\right)^{-\frac{5\sqrt{3}}{3}}$; 2) $x^{-\frac{\sqrt{3}}{5}} + \frac{1}{5} = 0$, $x_2 = \left(-\frac{1}{5}\right)^{-\frac{5\sqrt{3}}{3}}$;

3) $x^{-\frac{\sqrt{3}}{5}} - \frac{2}{5} = 0$, $x_3 = \left(\frac{2}{5}\right)^{-\frac{5\sqrt{3}}{3}}$; 4) $x^{-\frac{\sqrt{3}}{5}} + \frac{2}{5} = 0$, $x_4 = \left(-\frac{2}{5}\right)^{-\frac{5\sqrt{3}}{3}}$;

5) $x^{-\frac{\sqrt{3}}{5}} - 10 = 0$, $x_5 = 10^{-\frac{5\sqrt{3}}{3}}$.

③: A Dinbakish Equation

$$ax^{\frac{8\sqrt{5}}{3}} + bx^{2\sqrt{5}} + cx^{\frac{4\sqrt{5}}{3}} + dx^{\frac{2\sqrt{5}}{3}} + e = 0,$$

$$(a \in R, b \in R, c \in R, d \in R, e \in R, a \neq 0, b \neq 0, c \neq 0, d \neq 0, e \neq 0)$$

One of the forms of the Dinbakish Equation can be:

$$\left(ax^{\frac{2\sqrt{5}}{3}} - 2\right)\left(x^{\frac{2\sqrt{5}}{3}} - 3a\right)\left(x^{\frac{2\sqrt{5}}{3}} + \frac{2}{3a}\right)\left(x^{\frac{2\sqrt{5}}{3}} + 3a\right) = 0$$

It expands:

$$ax^{\frac{8\sqrt{5}}{3}} - \frac{4}{3}x^{2\sqrt{5}} - \left(9a^3 + \frac{4}{3a}\right)x^{\frac{4\sqrt{5}}{3}} + 12a^2x^{\frac{2\sqrt{5}}{3}} + 12a = 0,$$

14

and $b = -\dfrac{4}{3}$, $c = -\left(9a^3 + \dfrac{4}{3a}\right)$, $d = 12a^2$, $e = 12a$.

The Dinbakish Equation follows the real number 2 to be changed.

When $a = \dfrac{1}{3}$, it means that:

$$c = -\dfrac{13}{3}, \quad d = \dfrac{4}{3}, \quad e = 4.$$

Now, the Dinbakish Equation is:

$$\dfrac{1}{3} x^{\frac{8\sqrt{5}}{3}} - \dfrac{4}{3} x^{2\sqrt{5}} - \dfrac{13}{3} x^{\frac{4\sqrt{5}}{3}} + \dfrac{4}{3} x^{\frac{2\sqrt{5}}{3}} + 4 = 0,$$

Its roots are: 1) $\dfrac{1}{3} x^{\frac{2\sqrt{5}}{3}} - 2 = 0$, $x_1 = 6^{\frac{3\sqrt{5}}{10}}$; 2) $x^{\frac{2\sqrt{5}}{3}} - 1 = 0$, $x_2 = 1$;

3) $x^{\frac{2\sqrt{5}}{3}} + 2 = 0$, $x_3 = (-2)^{\frac{3\sqrt{5}}{10}}$; 4) $x^{\frac{2\sqrt{5}}{3}} + 1 = 0$, $x_4 = (-1)^{\frac{3\sqrt{5}}{10}}$.

④ : A Dinbakish Equations

$$ax^{-6\sqrt{6}} + bx^{-\frac{9\sqrt{6}}{2}} + cx^{-3\sqrt{6}} + dx^{-\frac{3\sqrt{6}}{2}} + e = 0,$$

$$\left(a \in R, b \in R, c \in R, d \in R, e \in R, a \neq 0, b \neq 0, c \neq 0, d \neq 0, e \neq 0\right)$$

One of the forms of the Dinbakish Equation can be:

$$\left(ax^{-\frac{3\sqrt{6}}{2}}-\frac{1}{4}\right)\left(x^{-\frac{3\sqrt{6}}{2}}+4a\right)\left(x^{-\frac{3\sqrt{6}}{2}}-4a\right)\left(x^{-\frac{3\sqrt{6}}{2}}+\frac{3}{4a}\right)=0,$$

It expands:

$$ax^{-6\sqrt{6}}+\frac{1}{2}x^{-\frac{9\sqrt{6}}{2}}-\left(16a^3+\frac{3}{16a}\right)x^{-3\sqrt{6}}-8a^2x^{-\frac{3\sqrt{6}}{2}}+3a=0,$$

and $b=\frac{1}{2}$, $c=-\left(16a^3+\frac{3}{16a}\right)$, $d=-8a^2$, $e=3a$.

The Dinbakish Equation follows the real number a to be changed.

When $a=\frac{1}{4}$, it means that:

$$c=-1,\ d=-\frac{1}{2},\ e=\frac{3}{4}.$$

Now, the Dinbakish Equation is:

$$\frac{1}{4}x^{-6\sqrt{6}}+\frac{1}{2}x^{-\frac{9\sqrt{6}}{2}}-x^{-3\sqrt{6}}-\frac{1}{2}x^{-\frac{3\sqrt{6}}{2}}+\frac{3}{4}=0,$$

Its roots are: 1) $\frac{1}{4}x^{-\frac{3\sqrt{6}}{2}} - \frac{1}{4} = 0$, $x_1 = 1$; 2) $x^{-\frac{3\sqrt{6}}{2}} + 1 = 0$, $x_2 = (-1)^{-\frac{\sqrt{6}}{9}}$;

3) $x^{-\frac{3\sqrt{6}}{2}} - 1 = 0$, $x_3 = 1$; 4) $x^{-\frac{3\sqrt{6}}{2}} + 3 = 0$, $x_4 = (-3)^{-\frac{\sqrt{6}}{9}}$.

4   Well, when the index numbers of the equation are irrational, not the irrational fractions, they are the Dinbakish Equations too.

I: $a_n x^{(n)\sqrt{k}} + a_{(n-1)} x^{(n-1)\sqrt{k}} + a_{(n-2)} x^{(n-2)\sqrt{k}} + \ldots + a_2 x^{(2)\sqrt{k}} + a_1 x^{\sqrt{k}} + a_0 = 0$;

II: $a_n x^{(n)(-\sqrt{k})} + a_{(n-1)} x^{(n-1)(-\sqrt{k})} + a_{(n-2)} x^{(n-2)(-\sqrt{k})} + \ldots + a_2 x^{(2)(-\sqrt{k})} + a_1 x^{-\sqrt{k}} + a_0 = 0$;

III: $a_n x^{(n)j\sqrt{k}} + a_{(n-1)} x^{(n-1)j\sqrt{k}} + a_{(n-2)} x^{(n-2)j\sqrt{k}} + \ldots + a_2 x^{(2)j\sqrt{k}} + a_1 x^{j\sqrt{k}} + a_0 = 0$;

IV: $a_n x^{(n)(-j\sqrt{k})} + a_{(n-1)} x^{(n-1)(-j\sqrt{k})} + a_{(n-2)} x^{(n-2)(-j\sqrt{k})} + \ldots + a_2 x^{(2)(-j\sqrt{k})} + a_1 x^{-j\sqrt{k}} + a_0 = 0$;

$$j = 2,3,4,\ldots,\infty; \quad k = 2,3,4,\ldots,\infty; \quad n = 1,2,3,\ldots,\infty.$$

①: A Dinbakish Equation

$$ax^{5\sqrt{5}} + bx^{4\sqrt{5}} + cx^{3\sqrt{5}} + dx^{2\sqrt{5}} + ex^{\sqrt{5}} + f = 0,$$

$$\left(a \in R, b \in R, c \in R, d \in R, e \in R, f \in R, a \neq 0, b \neq 0, c \neq 0, d \neq 0, e \neq 0, f \neq 0\right)$$

One of the forms of the Dinbakish Equation can be:

$$\left(ax^{\sqrt5}-\frac{2}{5a}\right)\left(x^{\sqrt5}+\frac{5a}{2}\right)\left(x^{\sqrt5}-\frac{3}{4a}\right)\left(x^{\sqrt5}+\frac{3}{4a}\right)\left(x^{\sqrt5}-4a\right)=0,$$

It expands:

$$ax^{5\sqrt5}-\left(\frac{3a^2}{2}+\frac{2}{5a}\right)x^{4\sqrt5}-\left(\frac{9}{16a}+10a^3-\frac{3}{5}\right)x^{3\sqrt5}+\left(4a+\frac{27}{32}+\frac{9}{40a^3}\right)x^{2\sqrt5}+\left(\frac{45a}{8}-\frac{27}{80a^2}\right)x^{\sqrt5}-\frac{9}{4a}=0,$$

and $b=-\left(\dfrac{3a^2}{2}+\dfrac{2}{5a}\right)$, $c=-\left(\dfrac{9}{16a}+10a^3-\dfrac{3}{5}\right)$, $d=4a+\dfrac{27}{32}+\dfrac{9}{40a^3}$,

$$e=\frac{45a}{8}-\frac{27}{80a^2},\quad f=-\frac{9}{4a}.$$

The Dinbakish Equation follows the real number a to be changed.

When $a=3$, it means that:

$$b=-\frac{409}{30},\quad c=-\frac{21567}{80},\quad d=\frac{6169}{480},\quad e=\frac{1347}{80},\quad f=-\frac{3}{4}.$$

Now, the Dinbakish Equation is:

$$3x^{5\sqrt5}-\frac{409}{30}x^{4\sqrt5}-\frac{21567}{80}x^{3\sqrt5}+\frac{6169}{480}x^{2\sqrt5}+\frac{1347}{80}x^{\sqrt5}-\frac{3}{4}=0.$$

18

Its roots are: 1) $x^{\sqrt{5}} - \dfrac{2}{45} = 0$, $x_1 = \left(\dfrac{2}{45}\right)^{\frac{\sqrt{5}}{5}}$; 2) $x^{\sqrt{5}} + \dfrac{15}{2} = 0$, $x_2 = \left(-\dfrac{15}{2}\right)^{\frac{\sqrt{5}}{5}}$;

3) $x^{\sqrt{5}} - \dfrac{1}{4} = 0$, $x_3 = \left(\dfrac{1}{4}\right)^{\frac{\sqrt{5}}{5}}$; 4) $x^{\sqrt{5}} + \dfrac{1}{4} = 0$, $x_4 = \left(-\dfrac{1}{4}\right)^{\frac{\sqrt{5}}{5}}$;

5) $x^{\sqrt{5}} - 12 = 0$, $x_5 = (12)^{\frac{\sqrt{5}}{5}}$.

②: A Dinbakish Equation

$$ax^{-5\sqrt{7}} + bx^{-4\sqrt{7}} + cx^{-3\sqrt{7}} + dx^{-2\sqrt{7}} + ex^{-\sqrt{7}} + f = 0$$

$$\left(a \in R, b \in R, c \in R, d \in R, e \in R, f \in R, a \neq 0, b \neq 0, c \neq 0, d \neq 0, e \neq 0, f \neq 0\right)$$

One of the forms of the Dinbakish Equation can be:

$$\left(ax^{-\sqrt{7}} - 3\right)\left(x^{-\sqrt{7}} - \dfrac{2}{5a}\right)\left(x^{-\sqrt{7}} + \dfrac{2}{5a}\right)\left(x^{-\sqrt{7}} - 5a\right)\left(x^{-\sqrt{7}} + 5a\right) = 0,$$

It expands:

$$ax^{-5\sqrt{7}} - 3x^{-4\sqrt{7}} - \left(25a^3 + \dfrac{4}{25a}\right)x^{-3\sqrt{7}} + \left(75a^2 + \dfrac{12}{25a^2}\right)x^{-2\sqrt{7}} + 4ax^{-\sqrt{7}} - 12 = 0,$$

and $b = -3$, $c = -\left(25a^3 + \dfrac{4}{25a}\right)$, $d = 75a^2 + \dfrac{12}{25a^2}$, $e = 4a$, $f = -12$.

The Dinbakish Equation follows the real number a to be changed.

When $a = -5$, it means that:

$$c = \frac{390629}{125}, \quad d = \frac{1171887}{625}, \quad e = -20.$$

Now, the Dinbakish Equation is:

$$-5x^{-5\sqrt{7}} - 3x^{-4\sqrt{7}} + \frac{390629}{125}x^{-3\sqrt{7}} + 1171887625\,x^{-2\sqrt{7}} - 20x^{-\sqrt{7}} - 12 = 0,$$

Its roots are: 1) $x^{-\sqrt{7}} + \dfrac{3}{5} = 0$, $x_1 = \left(-\dfrac{3}{5}\right)^{-\frac{\sqrt{7}}{7}}$; 2) $x^{-\sqrt{7}} + \dfrac{2}{25} = 0$, $x_2 = \left(-\dfrac{2}{25}\right)^{-\frac{\sqrt{7}}{7}}$;

3) $x^{-\sqrt{7}} - \dfrac{2}{25} = 0$, $x_3 = \left(\dfrac{2}{25}\right)^{-\frac{\sqrt{7}}{7}}$; 4) $x^{-\sqrt{7}} + 25 = 0$, $x_4 = (-25)^{-\frac{\sqrt{7}}{7}}$;

5) $x^{-\sqrt{7}} - 25 = 0$, $x_5 = (25)^{-\frac{\sqrt{7}}{7}}$.

5 The Composite Dinbakish Equations

Looking at this:

$$\left(2x^{\frac{1}{3}}-1\right)\left(x^{\frac{1}{3}}+1\right)\left(x^{\frac{1}{3}}-2\right)\left(x^{\frac{1}{3}}+2\right)\left(x^{\frac{\sqrt{3}}{3}}-3\right)=0 ,$$

It expands:

$$2x^{\frac{4+\sqrt{3}}{3}}-6x^{\frac{4}{3}}+x^{\frac{3+\sqrt{3}}{3}}-3x-9x^{\frac{2+\sqrt{3}}{3}}+27x^{\frac{2}{3}}-4x^{\frac{1+\sqrt{3}}{3}}+12x^{\frac{1}{3}}+4x^{\frac{\sqrt{3}}{3}}-12=0 ,$$

Now, let the coefficient number 2 to be a_{10}, the coefficient number -6 to be a_9, the coefficient number 1 to be a_8, the coefficient number -3 to be a_7, the coefficient number -9 to be a_6, the coefficient number 27 to be a_5, the coefficient number -4 to be a_4, the coefficient number 12 to be a_3, the coefficient number 4 to be a_2, and the integer number -12 to be a_1.

Now, it is $a_{10}x^{\frac{4+\sqrt{3}}{3}}+a_9x^{\frac{4}{3}}+a_8x^{\frac{3+\sqrt{3}}{3}}+a_7x+a_6x^{\frac{2+\sqrt{3}}{3}}+a_5x^{\frac{2}{3}}+a_4x^{\frac{1+\sqrt{3}}{3}}+a_3x^{\frac{1}{3}}+a_2x^{\frac{\sqrt{3}}{3}}+a_1=0 .$

In an equation, its index numbers have two or more than two series of different Dinbakish Equations to be composed together, I call it the Composite Dinbakish Equation.

❶ : Ones of the general forms of the Composite Dinbakish Equations are:

i: $a_{2n}x^{(n-1)\left(\frac{1}{k}\right)+\frac{\sqrt{k}}{k}}+a_{2n-1}x^{(n-1)\frac{1}{k}}+a_{2n-2}x^{(n-2)\left(\frac{1}{k}\right)+\frac{\sqrt{k}}{k}}+a_{2n-3}x^{(n-2)\frac{1}{k}}+...+a_3x^{\frac{1}{k}}+a_2x^{\frac{\sqrt{k}}{k}}+a_1=0$;

ii: $a_{2n}x^{(n-1)\left(\frac{1}{k}\right)+\frac{\sqrt{j}}{k}}+a_{2n-1}x^{(n-1)\frac{1}{k}}+a_{2n-2}x^{(n-2)\left(\frac{1}{k}\right)+\frac{\sqrt{j}}{k}}+a_{2n-3}x^{(n-2)\frac{1}{k}}+...+a_3x^{\frac{1}{k}}+a_2x^{\frac{\sqrt{j}}{k}}+a_1=0$.

($a_{2n}\neq0; a_{2n}\in R; n=2,3,4,5,...,\infty; k=2,3,4,5,...,\infty; j=2,3,4,5,...,\infty$)

(1) : A Composite Dinbakish Equation

$$a_8x^{\frac{3+\sqrt{5}}{2}}+a_7x^2+a_6x^{\frac{2+\sqrt{5}}{2}}+a_5x+a_4x^{\frac{1+\sqrt{5}}{2}}+a_3x^2+a_2x^{\frac{\sqrt{5}}{2}}+a_1=0 ,$$

One of the forms of this Composite Dinbakish Equation can be:

$$\left(a_8x^{\frac{1}{2}}-2\right)\left(x^{\frac{1}{2}}-\frac{a_8}{2}\right)\left(x^{\frac{1}{2}}-\frac{1}{a_8}\right)\left(x^{\frac{\sqrt{5}}{2}}+\frac{1}{a_8}\right)=0 ,$$

21

It expands:

$$a_8 x^{\frac{3+\sqrt{5}}{2}} + x^{\frac{3}{2}} - (3+\frac{a_8^2}{2})x^{\frac{2+\sqrt{5}}{2}} - \frac{1}{a_8}(3+\frac{a_8^2}{2})x + (\frac{3a_8}{2}+\frac{2}{a_8})x^{\frac{1+\sqrt{5}}{2}} + (\frac{3}{2}+\frac{2}{a_8^2})x^{\frac{1}{2}} - x^{\frac{\sqrt{5}}{2}} - \frac{1}{a_8} = 0 \ ,$$

And $a_7 = 1$, $a_6 = -(3+\frac{a_8^2}{2})$, $a_5 = -\frac{1}{a_8}(3+\frac{a_8^2}{2})$, $a_4 = \frac{3a_8}{2}+\frac{2}{a_8}$, $a_3 = \frac{3}{2}+\frac{2}{a_8^2}$, $a_2 = -1$, $a_1 = -\frac{1}{a_8}$.

The Composite Dinbakish Equation follows the real number a_8 to be changed.

When $a_8 = -4$, it means that:

$$a_6 = -11; a_5 = \frac{11}{4}; a_4 = -\frac{13}{2}; a_3 = \frac{13}{8}; a_1 = \frac{1}{4} \ .$$

Now the Composite Dinbakish Equation is:

$$-4x^{\frac{3+\sqrt{5}}{2}} + x^{\frac{3}{2}} - 11x^{\frac{2+\sqrt{5}}{2}} + \frac{11}{4}x - \frac{13}{2}x^{\frac{1+\sqrt{5}}{2}} + \frac{13}{8}x^{\frac{1}{2}} - x^{\frac{\sqrt{5}}{2}} + \frac{1}{4} = 0 \ .$$

Its roots are: 1) $-4x^{\frac{1}{2}} - 2 = 0$, $x_1 = \left(-\frac{1}{2}\right)^2$;   2) $x^{\frac{1}{2}} + 2 = 0$, $x_2 = (-2)^2$;

3) $x^{\frac{1}{2}} + \frac{1}{4} = 0$, $x_3 = \left(-\frac{1}{4}\right)^2$;   4) $x^{\frac{\sqrt{5}}{2}} - \frac{1}{4} = 0$; $x_4 = \left(\frac{1}{4}\right)^{\frac{2\sqrt{5}}{5}}$.

(2) : A Composite Dinbakish Equation

$$a_6 x^{\frac{2+\sqrt{3}}{3}} + a_5 x^{\frac{2}{3}} + a_4 x^{\frac{1+\sqrt{3}}{3}} + a_3 x^{\frac{1}{3}} + a_2 x^{\frac{\sqrt{3}}{3}} + a_1 = 0 \ ,$$

One of the forms of this Composite Dinbakish Equation can be:

$$\left(a_6 x^{\frac{1}{3}} + 3\right)\left(x^{\frac{1}{3}} - 4a_6\right)\left(x^{\frac{\sqrt{3}}{3}} + 2a_6\right) = 0 \ .$$

22

It expands:

$$a_6 x^{\frac{2+\sqrt{3}}{3}} + 2a_6^2 x^{\frac{2}{3}} - \left(4a_6^2 - 3\right)x^{\frac{1+\sqrt{3}}{3}} - \left(8a_6^3 - 6a_6\right)x^{\frac{1}{3}} - 12a_6 x^{\frac{\sqrt{3}}{3}} - 24a_6^2 = 0 \ ,$$

And $a_5 = 2a_6^2$, $a_4 = -\left(4a_6^2 - 3\right)$, $a_3 = -\left(8a_6^3 - 6a_6\right)$, $a_2 = -12a_6$, $a_1 = -24a_6^2$.

The Composite Dinbakish Equation follows the real number a_6 to be changed.

When $a_6 = 7$, it means that:

$a_5 = 98; a_4 = -193; a_3 = -2702; a_2 = -84; a_1 = -1176$.

Now, the Composite Dinbakish Equation is:

$$7x^{\frac{2+\sqrt{3}}{3}} + 98x^{\frac{2}{3}} - 193x^{\frac{1+\sqrt{3}}{3}} - 2702x^{\frac{1}{3}} - 84x^{\frac{\sqrt{3}}{3}} - 1176 = 0 \ ,$$

Its roots are: 1) $7x^{\frac{1}{3}} + 3 = 0; x_1 = \left(-\frac{3}{7}\right)^3$; 2) $x^{\frac{1}{3}} - 28 = 0$, $x_2 = (28)^3$;

3) $x^{\frac{\sqrt{3}}{3}} + 14 = 0$, $x_3 = (-14)^{\sqrt{3}}$.

❷ : Ones of the general forms of the Composite Dinbakish Equations are:

i: $a_{2n}x^{(n-1)\left(\frac{1}{k}\right)+\frac{j\sqrt{k}}{k}} + a_{2n-1}x^{(n-1)\frac{1}{k}} + a_{2n-2}x^{(n-2)\left(\frac{1}{k}\right)+\frac{j\sqrt{k}}{k}} + a_{2n-3}x^{(n-2)\frac{1}{k}} + ... + a_3 x^{\frac{1}{k}} + a_2 x^{\frac{j\sqrt{k}}{k}} + a_1 = 0$;

ii: $a_{2n}x^{(n-1)\left(\frac{1}{k}\right)+\frac{j\sqrt{i}}{k}} + a_{2n-1}x^{(n-1)\frac{1}{k}} + a_{2n-2}x^{(n-2)\left(\frac{1}{k}\right)+\frac{j\sqrt{i}}{k}} + a_{2n-3}x^{(n-2)\frac{1}{k}} + ... + a_3 x^{\frac{1}{k}} + a_2 x^{\frac{j\sqrt{i}}{k}} + a_1 = 0$.

$$\left(i = 2,3,4,...,\infty; j = 2,3,4,...,\infty; k = 2,3,4,...,\infty; n = 2,3,4,...,\infty\right)$$

$\boxed{(1)}$: A Composite Dinbakish Equation

$$a_6 x^{\frac{2+3\sqrt{5}}{5}} + a_5 x^{\frac{2}{5}} + a_4 x^{\frac{1+3\sqrt{5}}{5}} + a_3 x^{\frac{1}{5}} + a_2 x^{\frac{3\sqrt{5}}{5}} + a_1 = 0 ,$$

One of the forms of this Composite Dinbakish Equation can be:

$$\left(a_6 x^{\frac{1}{5}} + 4 \right)\left(x^{\frac{1}{5}} + 3a_6 \right)\left(x^{\frac{3\sqrt{5}}{5}} - \frac{1}{3a_6} \right) = 0 .$$

It expands:

$$a_6 x^{\frac{2+3\sqrt{5}}{5}} - \frac{1}{3} x^{\frac{2}{5}} + \left(3a_6^2 + 4 \right) x^{\frac{1+3\sqrt{5}}{5}} - \left(a_6 + \frac{4}{3a_6} \right) x^{\frac{1}{5}} + 12a_6 x^{\frac{3\sqrt{5}}{5}} - 4 = 0 ,$$

And $a_5 = -\dfrac{1}{3}$, $a_4 = 3a_6^2 + 4$, $a_3 = -(a_6 + \dfrac{4}{3a_6})$, $a_2 = 12a_6$, $a_1 == -4$.

The Composite Dinbakish Equation follows the real number a_6 to be changed.

When $a_6 = -6$, it means that:

$$a_4 = 112 , \; a_3 = \frac{56}{9} , \; a_2 = -72 .$$

Now, the Composite Dinbakish Equation is:

$$-6x^{\frac{2+3\sqrt{5}}{5}} - \frac{1}{3} x^{\frac{2}{5}} + 112 x^{\frac{1+3\sqrt{5}}{5}} + \frac{56}{9} x^{\frac{1}{5}} - 72 x^{\frac{3\sqrt{5}}{5}} - 4 = 0 ,$$

Its roots are: 1) $-6x^{\frac{1}{5}} + 4 = 0$, $x_1 = \left(\dfrac{2}{3} \right)^5$; 2) $x^{\frac{1}{5}} - 18 = 0$, $x_2 = (18)^5$;

$$3) \; x^{\frac{3\sqrt{5}}{5}} + \frac{1}{18} = 0 , \; x_3 = \left(-\frac{1}{18} \right)^{\frac{\sqrt{5}}{3}} .$$

$$\left(2\right) : \text{ A Composite Dinbakish Equation}$$

$$a_8 x^{\frac{3+3\sqrt{2}}{4}} + a_7 x^{\frac{3}{4}} + a_6 x^{\frac{2+3\sqrt{2}}{4}} + a_5 x^{\frac{2}{4}} + a_4 x^{\frac{1+3\sqrt{2}}{4}} + a_3 x^{\frac{1}{4}} + a_2 x^{\frac{3\sqrt{2}}{4}} + a_1 = 0 \ .$$

One of the forms of this Composite Dinbakish Equation can be:

$$\left(a_8 x^{\frac{1}{4}} + 5 \right)\left(x^{\frac{1}{4}} - \frac{3}{a_8} \right)\left(x^{\frac{1}{4}} - \frac{a_8}{3} \right)\left(x^{\frac{3\sqrt{2}}{4}} - a_8 \right) = 0 \ ,$$

It expands:

$$a_8 x^{\frac{3+3\sqrt{2}}{4}} - a_8^2 x^{\frac{3}{4}} - \left(\frac{a_8^2}{3} - 2 \right) x^{\frac{2+3\sqrt{2}}{4}} + \left(\frac{a_8^3}{3} - 2a_8 \right) x^{\frac{2}{4}} - \left(\frac{2a_8}{3} + \frac{15}{a_8} \right) x^{\frac{1+3\sqrt{2}}{4}} + \left(\frac{2a_8^2}{3} + 15 \right) x^{\frac{1}{4}} + 5x^{\frac{3\sqrt{2}}{4}} - 5a_8 = 0 \ ,$$

And $a_7 = -a_8^2$, $a_6 = -\left(\dfrac{a_8^2}{3} - 2 \right)$, $a_5 = \dfrac{a_8^3}{3} - 2a_8$, $a_4 = -\left(\dfrac{2a_8}{3} + \dfrac{15}{a_8} \right)$, $a_3 = \dfrac{2a_8^2}{3} + 15$,

$$a_2 = 5 \ , \ a_1 = -5a_8 \ .$$

The Composite Dinbakish Equation follows the real number a_8 to be changed.

When $a_8 = 6$, it means that:

$$a_7 = -36 \ , \ a_6 = -10 \ , \ a_5 = 60 \ , \ a_4 = -\frac{13}{2} \ , \ a_3 = 39 \ , \ a_1 = -30 \ .$$

Now, the Composite Dinbakish Equation is:

$$6x^{\frac{3+3\sqrt{2}}{4}} - 36x^{\frac{3}{4}} - 10x^{\frac{2+3\sqrt{2}}{4}} + 60x^{\frac{2}{4}} - \frac{13}{2} x^{\frac{1+3\sqrt{2}}{4}} + 39x^{\frac{1}{4}} + 5x^{\frac{3\sqrt{2}}{4}} - 30 = 0 \ ,$$

Its roots are: 1) $6x^{\frac{1}{4}} + 5 = 0$, $x_1 = \left(-\dfrac{5}{6} \right)^4$; 2) $x^{\frac{1}{4}} - \dfrac{1}{2} = 0$, $x_2 = \left(\dfrac{1}{2} \right)^4$;

3) $x^{\frac{1}{4}} - 2 = 0$, $x_3 = 2^4$; 4) $x^{\frac{3\sqrt{2}}{4}} - 6 = 0$, $x_4 = 6^{\frac{2\sqrt{2}}{3}}$.

3 : Ones of the general forms of the Composite Dinbakish Equations

i: $a_{2n}x^{(n-1)\left(\frac{i}{k}\right)+\frac{\sqrt{k}}{k}} + a_{2n-1}x^{(n-1)\frac{i}{k}} + a_{2n-2}x^{(n-2)\left(\frac{i}{k}\right)+\frac{\sqrt{k}}{k}} + a_{2n-3}x^{(n-2)\frac{i}{k}} + ... + a_3x^{\frac{i}{k}} + a_2x^{\frac{\sqrt{k}}{k}} + a_1 = 0$;

ii: $a_{2n}x^{(n-1)\left(\frac{i}{k}\right)+\frac{j\sqrt{k}}{k}} + a_{2n-1}x^{(n-1)\frac{i}{k}} + a_{2n-2}x^{(n-2)\left(\frac{i}{k}\right)+\frac{j\sqrt{k}}{k}} + a_{2n-3}x^{(n-2)\frac{i}{k}} + ... + a_3x^{\frac{i}{k}} + a_2x^{\frac{j\sqrt{k}}{k}} + a_1 = 0$;

iii: $a_{2n}x^{(n-1)\left(\frac{i}{k}\right)+\frac{\sqrt{g}}{k}} + a_{2n-1}x^{(n-1)\frac{i}{k}} + a_{2n-2}x^{(n-2)\left(\frac{i}{k}\right)+\frac{\sqrt{g}}{k}} + a_{2n-3}x^{(n-2)\frac{i}{k}} + ... + a_3x^{\frac{i}{k}} + a_2x^{\frac{\sqrt{g}}{k}} + a_1 = 0$;

iv: $a_{2n}x^{(n-1)\left(\frac{i}{k}\right)+\frac{j\sqrt{g}}{k}} + a_{2n-1}x^{(n-1)\frac{i}{k}} + a_{2n-2}x^{(n-2)\left(\frac{i}{k}\right)+\frac{j\sqrt{g}}{k}} + a_{2n-3}x^{(n-2)\frac{i}{k}} + ... + a_3x^{\frac{i}{k}} + a_2x^{\frac{j\sqrt{g}}{k}} + a_1 = 0$.

$$\left(n = 2,3,4,...,\infty; i = 2,3,4,...,\infty; k = 2,3,4,...,\infty; g = 2,3,4,...,\infty\right)$$

(1) :   A Composite Dinbakish Equation

$$a_6x^{\frac{6+\sqrt{5}}{5}} + a_5x^{\frac{6}{5}} + a_4x^{\frac{3+\sqrt{5}}{5}} + a_3x^{\frac{3}{5}} + a_2x^{\frac{\sqrt{5}}{5}} + a_1 = 0$$

One of the forms of this Composite Dinbakish Equation can be:

$$\left(a_6x^{\frac{3}{5}} + 7\right)\left(x^{\frac{3}{5}} - \frac{2}{a_6}\right)\left(x^{\frac{\sqrt{5}}{5}} + \frac{2}{a_6}\right) = 0$$;

It expands:

$$a_6x^{\frac{6+\sqrt{5}}{5}} + 2x^{\frac{6}{5}} + 5x^{\frac{3+\sqrt{5}}{5}} + \frac{10}{a_6}x^{\frac{3}{5}} - \frac{14}{a_6}x^{\frac{\sqrt{5}}{5}} - \frac{28}{a_6^2} = 0$$,

And $a_5 = 2; a_4 = 5; a_3 = \dfrac{10}{a_6}; a_2 = -\dfrac{14}{a_6}; a_1 = -\dfrac{28}{a_6^2}$.

The Composite Dinbakish Equation follows the real number a_6 to be changed.

When $a_6 = \dfrac{1}{2}$, it means that:

$$a_3 = 20 \ , \ a_2 = -28 \ , \ a_1 = -112 \ .$$

Now, the Composite Dinbakish Equation is:

$$\frac{1}{2}x^{\frac{6+\sqrt{5}}{5}} + 2x^{\frac{6}{5}} + 5x^{\frac{3+\sqrt{5}}{5}} + 20x^{\frac{3}{5}} - 28x^{\frac{\sqrt{5}}{5}} - 112 = 0 \ ,$$

Its roots are: 1) $\dfrac{1}{2}x^{\frac{3}{5}} + 7 = 0$, $x_1 = (-14)^{\frac{5}{3}}$; 2) $x^{\frac{3}{5}} - 4 = 0$, $x_2 = 4^{\frac{5}{3}}$;

$$3) \ x^{\frac{\sqrt{5}}{5}} + 4 = 0 , \ x_3 = (-4)^{\sqrt{5}} \ .$$

(2) :   A Composite Dinbakish Equation

$$a_8 x^{\frac{9+5\sqrt{2}}{2}} + a_7 x^{\frac{9}{2}} + a_6 x^{\frac{6+5\sqrt{2}}{2}} + a_5 x^3 + a_4 x^{\frac{3+5\sqrt{2}}{2}} + a_3 x^{\frac{3}{2}} + a_2 x^{\frac{5\sqrt{2}}{2}} + a_1 = 0$$

One of the forms of this Composite Dinbakish Equation can be:

$$\left(a_8 x^{\frac{3}{2}} - 7\right)\left(x^{\frac{3}{2}} - \frac{2}{3a_8}\right)\left(x^{\frac{3}{2}} + \frac{2}{3a_8}\right)\left(x^{\frac{5\sqrt{2}}{2}} - 3\right) = 0 \ ,$$

It expands:

$$a_8 x^{\frac{9+5\sqrt{2}}{2}} - 3a_8 x^{\frac{9}{2}} - 7x^{\frac{6+5\sqrt{2}}{2}} + 21x^3 - \frac{4}{9a_8}x^{\frac{3+5\sqrt{2}}{2}} + \frac{4}{3a_8}x^{\frac{3}{2}} + \frac{28}{9a_8^2}x^{\frac{5\sqrt{2}}{2}} - \frac{28}{3a_8^2} = 0 \ ,$$

And $a_7 = -3a_8; a_6 = -7; a_5 = 21; a_4 = -\dfrac{4}{9a_8}; a_3 = \dfrac{4}{3a_8}; a_2 = \dfrac{28}{9a_8^2}; a_1 = -\dfrac{28}{3a_8^2}$.

The Composite Dinbakish Equation follows the real number a_8 to be changed.

When $a_8 = \dfrac{1}{3}$, it means that:

$$a_7 = -1 \;,\; a_4 = -\frac{4}{3} \;,\; a_3 = 4 \;,\; a_2 = 28 \;,\; a_1 = -84 \;.$$

Now, the Composite Dinbakish Equation is:

$$\frac{1}{3}x^{\frac{9+5\sqrt{2}}{2}} - x^{\frac{9}{2}} - 7x^{\frac{6+5\sqrt{2}}{2}} + 21x^3 - \frac{4}{3}x^{\frac{3+5\sqrt{2}}{2}} + 4x^{\frac{3}{2}} + 28x^{\frac{5\sqrt{2}}{2}} - 84 = 0 \;,$$

Its roots are: 1) $\frac{1}{3}x^{\frac{3}{2}} - 7 = 0$, $x_1 = (21)^{\frac{2}{3}}$; 2) $x^{\frac{3}{2}} - 2 = 0$, $x_2 = 2^{\frac{2}{3}}$;

2) $x^{\frac{3}{2}} + 2 = 0$, $x_3 = (-2)^{\frac{2}{3}}$;; 4) $x^{\frac{5\sqrt{2}}{2}} - 3 = 0$, $x_4 = 3^{\frac{\sqrt{2}}{5}}$.

$$(3) : \text{A Composite Dinbakish Equation}$$

$$a_6 x^{\frac{8+\sqrt{6}}{5}} + a_5 x^{\frac{8}{5}} + a_4 x^{\frac{4+\sqrt{6}}{5}} + a_3 x^{\frac{4}{5}} + a_2 x^{\frac{\sqrt{6}}{5}} + a_1 = 0$$

One of the forms of this Composite Dinbakish Equation can be:

$$\left(a_6 x^{\frac{4}{5}} + 8\right)\left(x^{\frac{4}{5}} - \frac{a_6}{2}\right)\left(x^{\frac{\sqrt{6}}{5}} - \frac{2}{a_6}\right) = 0 \;,$$

It expands:

$$a_6 x^{\frac{8+\sqrt{6}}{5}} - 2x^{\frac{8}{5}} - \left(\frac{a_6^2}{2} - 8\right)x^{\frac{4+\sqrt{6}}{5}} + \left(a_6 - \frac{16}{a_6}\right)x^{\frac{4}{5}} + 4a_6 x^{\frac{\sqrt{6}}{5}} + 8 = 0 \;,$$

And $a_5 = -2; a_4 = -\left(\frac{a_6^2}{2} - 8\right); a_3 = a_6 - \frac{16}{a_6}; a_2 = -4a_6, a_1 = 8$.

The Composite Dinbakish Equation follows the real number a_6 to be changed.

When $a_6 = 2$, it means that:

$$a_4 = 6; a_3 = -6; a_2 = -8 \;.$$

Now, the Composite Dinbakish Equation is:

$$2x^{\frac{8+\sqrt{6}}{5}} - 2x^{\frac{8}{5}} + 6x^{\frac{4+\sqrt{6}}{5}} - 6x^{\frac{4}{5}} - 8x^{\frac{\sqrt{6}}{5}} + 8 = 0 \; ,$$

Its roots are: 1) $2x^{\frac{4}{5}} + 8 = 0$, $x_1 = (-4)^{\frac{5}{4}}$; 2) $x^{\frac{4}{5}} - 1 = 0$, $x_2 = 1$;

3) $x^{\frac{\sqrt{6}}{5}} - 1 = 0$, $x_3 = 1$.

(4) : A Composite Dinbakish Equation

$$a_8 x^{\frac{15+3\sqrt{5}}{7}} + a_7 x^{\frac{15}{7}} + a_6 x^{\frac{10+3\sqrt{5}}{7}} + a_5 x^{\frac{10}{7}} + a_4 x^{\frac{5+3\sqrt{5}}{7}} + a_3 x^{\frac{5}{7}} + a_2 x^{\frac{3\sqrt{5}}{7}} + a_1 = 0$$

One of the forms of this Composite Dinbakish Equation can be:

$$\left(a_8 x^{\frac{5}{7}} - 9 \right)\left(x^{\frac{5}{7}} - \frac{3}{a_8} \right)\left(x^{\frac{5}{7}} + \frac{a_8}{3} \right)\left(x^{\frac{3\sqrt{5}}{7}} - 6 \right) = 0 \; ,$$

It expands:

$$a_8 x^{\frac{15+3\sqrt{5}}{7}} - 6a_8 x^{\frac{15}{7}} + \left(\frac{a_8^2}{3} - 12 \right) x^{\frac{10+3\sqrt{5}}{7}} - 6\left(\frac{a_8^2}{3} - 12 \right) x^{\frac{10}{7}} + \left(\frac{27}{a_8} - 4a_8 \right) x^{\frac{5+3\sqrt{5}}{7}} - 6\left(\frac{27}{a_8} - 4a_8 \right) x^{\frac{5}{7}} + a_8 x^{\frac{3\sqrt{5}}{7}} - 6a_8 = 0 \; ,$$

And $a_7 = -6a_8; a_6 = \frac{a_8^2}{3} - 12; a_5 = -6\left(\frac{a_8^2}{3} - 12 \right); a_4 = \frac{27}{a_8} - 4a_8; a_3 = -6\left(\frac{27}{a_8} - 4a_8 \right); a_2 = 9; a_1 = -54$.

The Composite Dinbakish Equation follows the real number a_8 to be changed.

When $a_8 = -3$, it means that:

$a_7 = 18; a_6 = -9; a_5 = 54; a_4 = 3; a_3 = -18;$.

Now, the Composite Dinbakish Equation is:

$$-3x^{\frac{15+3\sqrt{5}}{7}} + 18x^{\frac{15}{7}} - 9x^{\frac{10+3\sqrt{5}}{7}} + 54x^{\frac{10}{7}} + 3x^{\frac{5+3\sqrt{5}}{7}} - 18x^{\frac{5}{7}} + 9x^{\frac{3\sqrt{5}}{7}} - 54 = 0 \; ,$$

Its roots are: 1) $-3x^{\frac{5}{7}} - 9 = 0$, $x_1 = (-3)^{\frac{7}{5}}$; 2) $x^{\frac{5}{7}} + 1 = 0$, $x_2 = (-1)^{\frac{7}{5}}$;

3) $x^{\frac{5}{7}} - 1 = 0$, $x_3 = 1$; 4) $x^{\frac{3\sqrt{5}}{7}} - 6 = 0$, $x_4 = 6^{\frac{7\sqrt{5}}{15}}$.

④ : Ones of the general forms of the Composite Dinbakish Equations:

i: $a_{2n}x^{(n-1)\left(\frac{\sqrt{k}}{k}\right)+\frac{j}{k}} + a_{2n-1}x^{(n-1)\frac{\sqrt{k}}{k}} + a_{2n-2}x^{(n-2)\left(\frac{\sqrt{k}}{k}\right)+\frac{j}{k}} + a_{2n-3}x^{(n-2)\frac{\sqrt{k}}{k}} + ... + a_3 x^{\frac{\sqrt{k}}{k}} + a_2 x^{\frac{j}{k}} + a_1 = 0$;

ii: $a_{2n}x^{(n-1)\left(\frac{\sqrt{i}}{k}\right)+\frac{j}{k}} + a_{2n-1}x^{(n-1)\frac{\sqrt{i}}{k}} + a_{2n-2}x^{(n-2)\left(\frac{\sqrt{i}}{k}\right)+\frac{j}{k}} + a_{2n-3}x^{(n-2)\frac{\sqrt{i}}{k}} + ... + a_3 x^{\frac{\sqrt{i}}{k}} + a_2 x^{\frac{j}{k}} + a_1 = 0$;

iii: $a_{2n}x^{(n-1)\left(\frac{m\sqrt{k}}{k}\right)+\frac{j}{k}} + a_{2n-1}x^{(n-1)\frac{m\sqrt{k}}{k}} + a_{2n-2}x^{(n-2)\left(\frac{m\sqrt{k}}{k}\right)+\frac{j}{k}} + a_{2n-3}x^{(n-2)\frac{m\sqrt{k}}{k}} + ... + a_3 x^{\frac{m\sqrt{k}}{k}} + a_2 x^{\frac{j}{k}} + a_1 = 0$;

iv: $a_{2n}x^{(n-1)\left(\frac{m\sqrt{i}}{k}\right)+\frac{j}{k}} + a_{2n-1}x^{(n-1)\frac{m\sqrt{i}}{k}} + a_{2n-2}x^{(n-2)\left(\frac{m\sqrt{i}}{k}\right)+\frac{j}{k}} + a_{2n-3}x^{(n-2)\frac{m\sqrt{i}}{k}} + ... + a_3 x^{\frac{m\sqrt{i}}{k}} + a_2 x^{\frac{j}{k}} + a_1 = 0$.

$$\left(i = 2,3,4,...,\infty; \, j = 1,2,3,...,\infty; k = 2,3,4,...,\infty; m = 2,3,4,...,\infty; n = 2,3,4,...,\infty \right)$$

(1) : A Composite Dinbakish Equation

$$a_6 x^{\frac{2\sqrt{3}+1}{3}} + a_5 x^{\frac{2\sqrt{3}}{3}} + a_4 x^{\frac{\sqrt{3}+1}{3}} + a_3 x^{\frac{\sqrt{3}}{3}} + a_2 x^{\frac{1}{3}} + a_1 = 0$$

One of the forms of this Composite Dinbakish Equation can be:

$$\left(a_6 x^{\frac{\sqrt{3}}{3}} - 9 \right)\left(x^{\frac{\sqrt{3}}{3}} - \frac{5}{a_6} \right)\left(x^{\frac{1}{3}} - a_6 \right) = 0 ,$$

It expands:

$$a_6 x^{\frac{2\sqrt{3}+1}{3}} - a_6^2 x^{\frac{2\sqrt{3}}{3}} - 14 x^{\frac{\sqrt{3}+1}{3}} + 14 a_6 x^{\frac{\sqrt{3}}{3}} + \frac{45}{a_6} x^{\frac{1}{3}} - 45 = 0 ,$$

30

And $a_5 = -a_6^2; a_4 = -14; a_3 = 14a_6; a_2 = \dfrac{45}{a_6}; a_1 = -45$.

The Composite Dinbakish Equation follows the real number a_6 to be changed.

When $a_6 = -5$, it means that:

$a_5 = -25; a_3 = -70; a_2 = -9$.

Now, the Composite Dinbakish Equation is:

$$-5x^{\frac{2\sqrt{3}+1}{3}} - 25x^{\frac{2\sqrt{3}}{3}} - 14x^{\frac{\sqrt{3}+1}{3}} - 70x^{\frac{\sqrt{3}}{3}} - 9x^{\frac{1}{3}} - 45 = 0 ,$$

Its roots are: 1) $-5x^{\frac{\sqrt{3}}{3}} - 9 = 0$, $x_1 = \left(-\dfrac{9}{5}\right)^{\sqrt{3}}$; 2) $x^{\frac{\sqrt{3}}{3}} + 1 = 0$, $x_2 = (-1)^{\sqrt{3}}$;

$$3) \; x^{\frac{1}{3}} + 5 = 0 , \; x_3 = (-5)^3 .$$

(2) : A Composite Dinbakish Equation

$$a_8 x^{\frac{3\sqrt{5}+5}{4}} + a_7 x^{\frac{3\sqrt{5}}{4}} + a_6 x^{\frac{2\sqrt{5}+5}{4}} + a_5 x^{\frac{\sqrt{5}}{2}} + a_4 x^{\frac{\sqrt{5}+5}{4}} + a_3 x^{\frac{\sqrt{5}}{4}} + a_2 x^{\frac{5}{4}} + a_1 = 0$$

One of the forms of this Composite Dinbakish Equation can be:

$$\left(a_8 x^{\frac{\sqrt{5}}{4}} - 10\right)\left(x^{\frac{\sqrt{5}}{4}} + 2a_8\right)\left(x^{\frac{\sqrt{5}}{4}} - \dfrac{3}{2a_8}\right)\left(x^{\frac{5}{4}} - 9\right) = 0 ,$$

It expands:

$$a_8 x^{\frac{3\sqrt{5}+5}{4}} - 9a_8 x^{\frac{3\sqrt{5}}{4}} + \left(2a_8^2 - \dfrac{23}{2}\right)x^{\frac{2\sqrt{5}+5}{4}} - 9\left(2a_8^2 - \dfrac{23}{2}\right)x^{\frac{\sqrt{5}}{2}} - \left(23a_8 - \dfrac{15}{a_8}\right)x^{\frac{\sqrt{5}+5}{4}} + 9\left(23a_8 - \dfrac{15}{a_8}\right)x^{\frac{\sqrt{5}}{4}} + 30x^{\frac{5}{4}} - 270 = 0 ,$$

And

$$a_7 = -9a_8; a_6 = 2a_8^2 - \dfrac{23}{2}; a_5 = -9\left(2a_8^2 - \dfrac{23}{2}\right); a_4 = -\left(23a_8 - \dfrac{15}{a_8}\right); a_3 = 9\left(23a_8 - \dfrac{15}{a_8}\right); a_2 = 30; a_1 = -270 .$$

The Composite Dinbakish Equation follows the real number a_8 to be changed.

When $a_8 = 3$, it means that:

$$a_7 = -27; a_6 = \frac{13}{2}; a_5 = -\frac{117}{2}; a_4 = -64; a_3 = 576 .$$

Now, the Composite Dinbakish Equation is:

$$3x^{\frac{3\sqrt{5}+5}{4}} - 27x^{\frac{3\sqrt{5}}{4}} + \frac{13}{2}x^{\frac{2\sqrt{5}+5}{4}} - \frac{117}{2}x^{\frac{\sqrt{5}}{2}} - 64x^{\frac{\sqrt{5}+5}{4}} + 576x^{\frac{\sqrt{5}}{4}} + 30x^{\frac{5}{4}} - 270 = 0 ,$$

Its roots are: 1) $3x^{\frac{\sqrt{5}}{4}} - 10 = 0$, $x_1 = \left(\frac{10}{3}\right)^{\frac{4\sqrt{5}}{5}}$; 2) $x^{\frac{\sqrt{5}}{4}} + 6 = 0$, $x_2 = (-6)^{\frac{4\sqrt{5}}{5}}$;

3) $x^{\frac{\sqrt{5}}{4}} - \frac{1}{2} = 0$, $x_3 = \left(\frac{1}{2}\right)^{\frac{4\sqrt{5}}{5}}$; 4) $x^{\frac{5}{4}} - 9 = 0$, $x_4 = 9^{\frac{4}{5}}$.

(3) : A Composite Dinbakish Equation

$$a_6 x^{\frac{6\sqrt{3}+4}{7}} + a_5 x^{\frac{6\sqrt{3}}{7}} + a_4 x^{\frac{4\sqrt{3}+4}{7}} + a_3 x^{\frac{4\sqrt{3}}{7}} + a_2 x^{\frac{4}{7}} + a_1 = 0$$

One of the forms of this Composite Dinbakish Equation can be:

$$\left(a_6 x^{\frac{2\sqrt{3}}{7}} - 11\right)\left(x^{\frac{2\sqrt{3}}{7}} - \frac{3}{a_6}\right)\left(x^{\frac{4}{7}} + \frac{4a_6}{3}\right) = 0 ,$$

It expands:

$$a_6 x^{\frac{6\sqrt{3}+4}{7}} + \frac{4a_6^2}{3}x^{\frac{6\sqrt{3}}{7}} - \frac{47}{4}x^{\frac{4\sqrt{3}+4}{7}} - \frac{47a_6}{3}x^{\frac{4\sqrt{3}}{7}} + \frac{33}{4a_6}x^{\frac{4}{7}} + 11 = 0 ,$$

And $a_5 = \frac{4a_6^2}{3}; a_4 = -\frac{47}{4}; a_3 = -\frac{47a_6}{3}; a_2 = \frac{33}{4a_6}; a_1 = 11$.

32

The Composite Dinbakish Equation follows the real number a_6 to be changed.

When $a_6 = -3$, it means that:

$$a_5 = 12; a_3 = 47; a_2 = -\frac{11}{4} \ .$$

Now, the Composite Dinbakish Equation is:

$$-3x^{\frac{6\sqrt{3}+4}{7}} + 12x^{\frac{6\sqrt{3}}{7}} - \frac{47}{4}x^{\frac{4\sqrt{3}+4}{7}} + 47x^{\frac{4\sqrt{3}}{7}} - \frac{11}{4}x^{\frac{4}{7}} + 11 = 0 \ ,$$

Its roots are: 1) $-3x^{\frac{2\sqrt{3}}{7}} - 11 = 0$, $x_1 = \left(-\frac{11}{3}\right)^{\frac{7\sqrt{3}}{6}}$; 2) $x^{\frac{2\sqrt{3}}{7}} + \frac{1}{4} = 0$, $x_2 = \left(-\frac{1}{4}\right)^{\frac{7\sqrt{3}}{6}}$;

3) $x^{\frac{4}{7}} - 4 = 0$, $x_3 = 4^{\frac{7}{4}}$.

(4) : A Composite Dinbakish Equation

$$a_8 x^{\frac{15\sqrt{2}+3}{2}} + a_7 x^{\frac{15\sqrt{2}}{2}} + a_6 x^{\frac{10\sqrt{2}+3}{2}} + a_5 x^{5\sqrt{2}} + a_4 x^{\frac{5\sqrt{2}+3}{2}} + a_3 x^{5\sqrt{2}} + a_2 x^{\frac{3}{2}} + a_1 = 0$$

One of the forms of this Composite Dinbakish Equation can be:

$$\left(a_8 x^{\frac{5\sqrt{2}}{2}} - 1\right)\left(x^{\frac{5\sqrt{2}}{2}} - a_8\right)\left(x^{\frac{5\sqrt{2}}{2}} - 2a_8\right)\left(x^{\frac{3}{2}} - 7\right) = 0 \ ,$$

It expands:

$$a_8 x^{\frac{15\sqrt{2}+3}{2}} - 7a_8 x^{\frac{15\sqrt{2}}{2}} - \left(3a_8^2+1\right)x^{\frac{10\sqrt{2}+3}{2}} + 7\left(3a_8^2+1\right)x^{5\sqrt{2}} + \left(2a_8^3+3a_8\right)x^{\frac{5\sqrt{2}+3}{2}} - 7\left(2a_8^3+3a_8\right)x^{5\sqrt{2}} - 2a_8^2 x^{\frac{3}{2}} + 14a_8^2 = 0 \ ,$$

And

$$a_7 = -7a_8; a_6 = -\left(3a_8^2+1\right); a_5 = 7\left(3a_8^2+1\right); a_4 = 2a_8^3+3a_8; a_3 = -7\left(2a_8^3+3a_8\right); a_2 = -2a_8^2; a_1 = 14a_8^2 \ .$$

The Composite Dinbakish Equation follows the real number a_8 to be changed.

33

When $a_8 = -9$, it means that:

$$a_7 = 63; a_6 = -244; a_5 = 1708; a_4 = -1485; a_3 = 10395; a_2 = -162; a_1 = 1134 \ .$$

Now, the Composite Dinbakish Equation is:

$$-9x^{\frac{15\sqrt{2}+3}{2}} + 63x^{\frac{15\sqrt{2}}{2}} - 244x^{\frac{10\sqrt{2}+3}{2}} + 1708x^{5\sqrt{2}} - 1485x^{\frac{5\sqrt{2}+3}{2}} + 10395x^{5\sqrt{2}} - 162x^{\frac{3}{2}} + 1134 = 0 \ ,$$

Its roots are: 1) $-9x^{\frac{5\sqrt{2}}{2}} - 1 = 0$, $x_1 = \left(-\dfrac{1}{9}\right)^{\frac{\sqrt{2}}{5}}$; 2) $x^{\frac{5\sqrt{2}}{2}} + 9 = 0$, $x_2 = \left(-9\right)^{\frac{\sqrt{2}}{5}}$;

3) $x^{\frac{5\sqrt{2}}{2}} + 18 = 0$, $x_3 = \left(-18\right)^{\frac{\sqrt{2}}{5}}$; 4) $x^{\frac{3}{2}} - 7 = 0$, $x_4 = 7^{\frac{2}{3}}$.

⑤: Ones of the general forms of the Composite Dinbakish Equations

i: $a_{2n}x^{(n-1)\frac{i}{k}+\sqrt{j}} + a_{2n-1}x^{(n-1)\frac{i}{k}} + a_{2a-2}x^{(n-2)\frac{i}{k}+\sqrt{j}} + a_{2n-3}x^{(n-2)\frac{i}{k}} + ... + a_3 x^{\frac{i}{k}} + a_2 x^{\sqrt{j}} + a_1 = 0$;

ii: $a_{2n}x^{(n-1)\frac{i}{k}+m\sqrt{j}} + a_{2n-1}x^{(n-1)\frac{i}{k}} + a_{2a-2}x^{(n-2)\frac{i}{k}+m\sqrt{j}} + a_{2n-3}x^{(n-2)\frac{i}{k}} + ... + a_3 x^{\frac{i}{k}} + a_2 x^{m\sqrt{j}} + a_1 = 0$;

$$\left(n = 2,3,4,...,\infty; m = 2,3,4,...,\infty; k = 2,3,4,...,\infty; i = 1,2,3,4,...,\infty; j = 2,3,4,...,\infty\right)$$

(1) : A Composite Dinbakish Equation

$$a_6 x^{1+\sqrt{3}} + a_5 x + a_4 x^{\frac{1}{2}+\sqrt{3}} + a_3 x^{\frac{1}{2}} + a_2 x^{\sqrt{3}} + a_1 = 0$$

One of the forms of this Composite Dinbakish Equation can be:

$$\left(a_6 x^{\frac{1}{2}} - 12\right)\left(x^{\frac{1}{2}} - \dfrac{6}{a_6}\right)\left(x^{\sqrt{3}} + \dfrac{a_6}{3}\right) = 0 \ ,$$

It expands:

$$a_6 x^{1+\sqrt{3}} + \dfrac{a_6^2}{3}x - 18x^{\frac{1}{2}+\sqrt{3}} - 6a_6 x^{\frac{1}{2}} + \dfrac{72}{a_6}x^{\sqrt{3}} + 24 = 0 \ ,$$

And $a_5 = \dfrac{a_6^2}{3}; a_4 = -18; a_3 = -6a_6; a_2 = \dfrac{72}{a_6}; a_1 = 24$.

The Composite Dinbakish Equation follows the real number a_6 to be changed.

When $a_6 = -6$, it means that:

$$a_5 = 12; a_3 = 36; a_2 = -12 .$$

Now, the Composite Dinbakish Equation is:

$$-6x^{1+\sqrt{3}} + 12x - 18x^{\frac{1}{2}+\sqrt{3}} + 36x^{\frac{1}{2}} - 12x^{\sqrt{3}} + 24 = 0 ,$$

Its roots are: 1) $-6x^{\frac{1}{2}} - 12 = 0$, $x_1 = (-2)^2$; 2) $x^{\frac{1}{2}} + 1 = 0$, $x_2 = (-1)^2$;

$$3) \ x^{\sqrt{3}} - 2 = 0 \ , \ x_3 = 2^{\frac{\sqrt{3}}{3}} .$$

(2) : A Composite Dinbakish Equation

$$a_8 x^{5+3\sqrt{6}} + a_7 x^5 + a_6 x^{\frac{10}{3}+3\sqrt{6}} + a_5 x^{\frac{10}{3}} + a_4 x^{\frac{5}{3}+3\sqrt{6}} + a_3 x^{\frac{5}{3}} + a_2 x^{3\sqrt{6}} + a_1 = 0$$

One of the forms of this Composite Dinbakish Equation can be:

$$\left(a_8 x^{\frac{5}{3}} - 4 \right)\left(x^{\frac{5}{3}} - a_8 \right)\left(x^{\frac{5}{3}} + 2a_8 \right)\left(x^{3\sqrt{6}} - 6 \right) = 0 ,$$

It expands:

$$a_8 x^{5+3\sqrt{6}} - 6a_8 x^5 + \left(a_8^2 - 4 \right)x^{\frac{10}{3}+3\sqrt{6}} - 6\left(a_8^2 - 4 \right)x^{\frac{10}{3}} - \left(2a_8^3 + 4a_8 \right)x^{\frac{5}{3}+3\sqrt{6}} + 6\left(2a_8^3 + 4a_8 \right)x^{\frac{5}{3}} + 8a_8^2 x^{3\sqrt{6}} - 48a_8^2 = 0 ,$$

And $a_7 = -6a_8; a_6 = a_8^2 - 4; a_5 = -6\left(a_8^2 - 4 \right); a_4 = -\left(2a_8^3 + 4a_8 \right); a_3 = 6\left(2a_8^3 + 4a_8 \right); a_2 = 8a_8^2; a_1 = -48a_8^2$.

The Composite Dinbakish Equation follows the real number a_8 to be changed.

When $a_8 = -4$, it means that:

$$a_7 = 24; a_6 = 12; a_5 = -72; a_4 = 144; a_3 = -864; a_2 = 128; a_1 = -768 \ .$$

Now, the Composite Dinbakish Equation is:

$$-4x^{5+3\sqrt{6}} + 24x^5 + 12x^{\frac{10}{3}+3\sqrt{6}} - 72x^{\frac{10}{3}} + 144x^{\frac{5}{3}+3\sqrt{6}} - 864x^{\frac{5}{3}} + 128x^{3\sqrt{6}} - 768 = 0 \ ,$$

Its roots are: 1) $-4x^{\frac{5}{3}} - 4 = 0$, $x_1 = (-1)^{\frac{3}{5}}$; 2) $x^{\frac{5}{3}} + 4 = 0$, $x_2 = (-4)^{\frac{3}{5}}$;

$\quad\quad$ 3) $x^{\frac{5}{3}} - 8 = 0$, $x_3 = 8^{\frac{3}{5}}$; 4) $x^{3\sqrt{6}} - 6 = 0$, $x_4 = 6^{\frac{\sqrt{6}}{18}}$.

❻: Ones of the general forms of the Composite Dinbakish Equations

i: $a_{2n}x^{(n-1)\sqrt{j}+\frac{i}{k}} + a_{2n-1}x^{(n-1)\sqrt{j}} + a_{2n-2}x^{(n-2)\sqrt{j}+\frac{i}{k}} + a_{2n-3}x^{(n-2)\sqrt{j}} + ... + a_3x^{\sqrt{j}} + a_2x^{\frac{i}{k}} + a_1 = 0$;

ii: $a_{2n}x^{(n-1)m\sqrt{j}+\frac{i}{k}} + a_{2n-1}x^{(n-1)m\sqrt{j}} + a_{2n-2}x^{(n-2)m\sqrt{j}+\frac{i}{k}} + a_{2n-3}x^{(n-2)m\sqrt{j}} + ... + a_3x^{m\sqrt{j}} + a_2x^{\frac{i}{k}} + a_1 = 0$;

$$\left(i = 1, 2, 3, ..., \infty; m = 2, 3, 4, ..., \infty; n = 2, 3, 4, ...\infty; j = 2, 3, 4, ..., \infty\right)$$

$\quad\quad$ (1) : A Composite Dinbakish Equation

$$a_6x^{2\sqrt{5}+\frac{4}{7}} + a_5x^{2\sqrt{5}} + a_4x^{\sqrt{5}+\frac{4}{7}} + a_3x^{\sqrt{5}} + a_2x^{\frac{4}{7}} + a_1 = 0$$

One of the forms of this Composite Dinbakish Equation can be:

$$\left(a_6x^{\sqrt{5}} + 5\right)\left(x^{\sqrt{5}} + \frac{5a_6}{2}\right)\left(x^{\frac{4}{7}} - \frac{4}{5a_6}\right) = 0 \ ,$$

It expands:

$$a_6x^{2\sqrt{5}+\frac{4}{7}} - \frac{4}{5}x^{2\sqrt{5}} + \left(\frac{5a_6^2}{2} + 5\right)x^{\sqrt{5}+\frac{4}{7}} - \left(2a_6 + \frac{4}{a_6}\right)x^{\sqrt{5}} + \frac{25a_6}{2}x^{\frac{4}{7}} - 10 = 0 \ ,$$

And $a_5 = -\dfrac{4}{5}; a_4 = \dfrac{5a_6^2}{2} + 5; a_3 = -\left(2a_6 + \dfrac{4}{a_6}\right); a_2 = \dfrac{25a_6}{2}; a_1 = -10$.

The Composite Dinbakish Equation follows the real number a_6 to be changed.

When $a_6 = -2$, it means that:

$$a_4 = 15; a_3 = 6; a_2 = -25 .$$

Now, the Composite Dinbakish Equation is:

$$-2x^{2\sqrt{5}+\frac{4}{7}} - \dfrac{4}{5}x^{2\sqrt{5}} + 15x^{\sqrt{5}+\frac{4}{7}} + 6x^{\sqrt{5}} - 25x^{\frac{4}{7}} - 10 = 0 ,$$

Its roots are: 1) $-2x^{\sqrt{5}} + 5 = 0$, $x_1 = \left(\dfrac{5}{2}\right)^{\frac{\sqrt{5}}{5}}$; 2) $x^{\sqrt{5}} - 5 = 0$, $x_2 = 5^{\frac{\sqrt{5}}{5}}$

$$3)\ x^{\frac{4}{7}} + \dfrac{2}{5} = 0 \ ,\ x_3 = \left(-\dfrac{2}{5}\right)^{\frac{7}{4}} .$$

$\boxed{(2)}$: A Composite Dinbakish Equation

$$a_8 x^{12\sqrt{7}+\frac{1}{6}} + a_7 x^{12\sqrt{7}} + a_6 x^{8\sqrt{7}+\frac{1}{6}} + a_5 x^{8\sqrt{7}} + a_4 x^{4\sqrt{7}+\frac{1}{6}} + a_3 x^{4\sqrt{7}} + a_2 x^{\frac{1}{6}} + a_1 = 0$$

One of the forms of this Composite Dinbakish Equation can be:

$$\left(a_8 x^{4\sqrt{7}} + 5\right)\left(x^{4\sqrt{7}} - \dfrac{4}{a_8}\right)\left(x^{4\sqrt{7}} - \dfrac{a_8}{4}\right)\left(x^{\frac{1}{6}} - 2\right) = 0 ,$$

It expands:

$$a_8 x^{12\sqrt{7}+\frac{1}{6}} - 2a_8 x^{12\sqrt{7}} - \left(\dfrac{a_8^2}{4} - 1\right)x^{8\sqrt{7}+\frac{1}{6}} + 2\left(\dfrac{a_8^2}{4} - 1\right)x^{8\sqrt{7}} - \left(\dfrac{a_8}{4} + \dfrac{20}{a_8}\right)x^{4\sqrt{7}+\frac{1}{6}} + 2\left(\dfrac{a_8}{4} + \dfrac{20}{a_8}\right)x^{4\sqrt{7}} + 5x^{\frac{1}{6}} - 10 = 0 ,$$

And $a_7 = -2a_8; a_6 = -\left(\dfrac{a_8^2}{4} - 1\right); a_5 = 2\left(\dfrac{a_8^2}{4} - 1\right); a_4 = -\left(\dfrac{a_8}{4} + \dfrac{20}{a_8}\right); a_3 = 2\left(\dfrac{a_8}{4} + \dfrac{20}{a_8}\right); a_2 = 5; a_1 = -10$.

The Composite Dinbakish Equation follows the real number a_8 to be changed.

When $a_8 = 4$, it means that:

$$a_7 = -8; a_6 = -3; a_5 = 6; a_4 = -6; a_3 = 12 \ .$$

Now, the Composite Dinbakish Equation is:

$$4x^{12\sqrt{7}+\frac{1}{6}} - 8x^{12\sqrt{7}} - 3x^{8\sqrt{7}+\frac{1}{6}} + 6x^{8\sqrt{7}} - 6x^{4\sqrt{7}+\frac{1}{6}} + 12x^{4\sqrt{7}} + 5x^{\frac{1}{6}} - 10 = 0 \ ,$$

Its roots are: 1) $4x^{4\sqrt{7}} + 5 = 0$, $x_1 = \left(-\dfrac{5}{4}\right)^{\frac{\sqrt{7}}{28}}$; 2) $x^{4\sqrt{7}} - 1 = 0$, $x_2 = 1$;

3) $x^{4\sqrt{7}} - 1 = 0$, $x_3 = 1$; 4) $x^{\frac{1}{6}} - 2 = 0$, $x_4 = 2^6$.

7 : One of the general forms of the Composite Dinbakish Equations

$$a_{2n}x^{(n-1)\frac{i}{k}+\frac{j}{m}} + a_{2n-1}x^{(n-1)\frac{i}{k}} + a_{2n-2}x^{(n-2)\frac{i}{k}+\frac{j}{m}} + a_{2n-3}x^{(n-2)\frac{i}{k}} + ... + a_3 x^{\frac{i}{k}} + a_2 x^{\frac{j}{m}} + a_1 = 0$$

$$\left(n = 2,3,4,...,\infty; k = 2,3,4,...,\infty; m = 2,3,4,...,\infty; i = 1,2,3,...,\infty; j = 1,2,3,...,\infty\right)$$

(1) : A Composite Dinbakish Equation

$$a_6 x^{\frac{6}{5}+\frac{1}{4}} + a_5 x^{\frac{6}{5}} + a_4 x^{\frac{3}{5}+\frac{1}{4}} + a_3 x^{\frac{3}{5}} + a_2 x^{\frac{1}{4}} + a_1 = 0$$

One of the forms of this Composite Dinbakish Equation can be:

$$\left(a_6 x^{\frac{3}{5}} + 5\right)\left(x^{\frac{3}{5}} + \frac{3}{a_6}\right)\left(x^{\frac{1}{4}} + 7\right) = 0 \ ,$$

It expands:

$$a_6 x^{\frac{6}{5}+\frac{1}{4}} + 7a_6 x^{\frac{6}{5}} + 8x^{\frac{3}{5}+\frac{1}{4}} + 56x^{\frac{3}{5}} + \frac{15}{a_6} x^{\frac{1}{4}} + \frac{105}{a_6} = 0 \ ,$$

And $a_5 = 7a_6; a_4 = 8; a_3 = 56; a_2 = \dfrac{15}{a_6}; a_1 = \dfrac{105}{a_6}$.

The Composite Dinbakish Equation follows the real number a_6 to be changed.

When $a_6 = 5$, it means that:

$$a_5 = 35; a_2 = 3; a_1 = 21 \ .$$

Now, the Composite Dinbakish Equation is:

$$5x^{\frac{6}{5}+\frac{1}{4}} + 35x^{\frac{6}{5}} + 8x^{\frac{3}{5}+\frac{1}{4}} + 56x^{\frac{3}{5}} + 3x^{\frac{1}{4}} + 21 = 0 \ ,$$

Its roots are: 1) $5x^{\frac{3}{5}} + 5 = 0$, $x_1 = (-1)^{\frac{5}{3}}$; 2) $x^{\frac{3}{5}} + \dfrac{3}{5} = 0$, $x_2 = \left(-\dfrac{3}{5}\right)^{\frac{5}{3}}$;

$$3) \ x^{\frac{1}{4}} + 7 = 0 \ , \ x_3 = (-7)^4 \ .$$

⑧ : More of the general forms of the Composite Dinbakish Equations

[1]: $a_{2n} x^{(n-1)\frac{i}{k}-\frac{\sqrt{j}}{k}} + a_{2n-1} x^{(n-1)\frac{i}{k}} + a_{2n-2} x^{(n-2)\frac{i}{k}-\frac{\sqrt{j}}{k}} + a_{2n-3} x^{(n-2)\frac{i}{k}} + ... + a_3 x^{\frac{i}{k}} + a_2 x^{-\frac{\sqrt{j}}{k}} + a_1 = 0$;

[2]: $a_{2n} x^{(n-1)\frac{i}{k}-\frac{\sqrt{j}}{m}} + a_{2n-1} x^{(n-1)\frac{i}{m}} + a_{2n-2} x^{(n-2)\frac{i}{k}-\frac{\sqrt{j}}{m}} + a_{2n-3} x^{(n-2)\frac{i}{k}} + ... + a_3 x^{\frac{i}{k}} + a_2 x^{-\frac{\sqrt{j}}{m}} + a_1 = 0$;

[3]: $a_{2n} x^{(n-1)\frac{i}{k}-\frac{g\sqrt{j}}{k}} + a_{2n-1} x^{(n-1)\frac{i}{k}} + a_{2n-2} x^{(n-2)\frac{i}{k}-\frac{g\sqrt{j}}{k}} + a_{2n-3} x^{(n-2)\frac{i}{k}} + ... + a_3 x^{\frac{i}{k}} + a_2 x^{-\frac{g\sqrt{j}}{k}} + a_1 = 0$;

[4]: $a_{2n} x^{(n-1)\frac{i}{k}-\frac{g\sqrt{j}}{m}} + a_{2n-1} x^{(n-1)\frac{i}{k}} + a_{2n-2} x^{(n-2)\frac{i}{k}-\frac{g\sqrt{j}}{m}} + a_{2n-3} x^{(n-2)\frac{i}{k}} + ... + a_3 x^{\frac{i}{k}} + a_2 x^{-\frac{g\sqrt{j}}{m}} + a_1 = 0$;

[5]: $a_{2n} x^{(n-1)(-\frac{g\sqrt{j}}{m})+\frac{i}{k}} + a_{2n-1} x^{(n-1)(-\frac{g\sqrt{j}}{m})} + a_{2n-2} x^{(n-2)(-\frac{g\sqrt{j}}{m})+\frac{i}{k}} + a_{2n-3} x^{(n-2)(-\frac{g\sqrt{j}}{m})} + ... + a_3 x^{-\frac{g\sqrt{j}}{m}} + a_2 x^{\frac{i}{k}} + a_1 = 0$;

[6]: $a_{2n}x^{(n-1)(-\frac{g\sqrt{j}}{k})+\frac{i}{k}} + a_{2n-1}x^{(n-1)(-\frac{g\sqrt{j}}{k})} + a_{2n-2}x^{(n-2)(-\frac{g\sqrt{j}}{k})+\frac{i}{k}} + a_{2n-3}x^{(n-2)(-\frac{g\sqrt{j}}{k})} + ... + a_{3}x^{-\frac{g\sqrt{j}}{k}} + a_{2}x^{\frac{i}{k}} + a_{1} = 0$;

[7]: $a_{2n}x^{(n-1)\frac{i}{k}-\sqrt{j}} + a_{2n-1}x^{(n-1)\frac{i}{k}} + a_{2n-2}x^{(n-2)\frac{i}{k}-\sqrt{j}} + a_{2n-3}x^{(n-2)\frac{i}{k}} + ... + a_{3}x^{\frac{i}{k}} + a_{2}x^{-\sqrt{j}} + a_{1} = 0$;

[8]: $a_{2n}x^{(n-1)\frac{i}{k}-g\sqrt{j}} + a_{2n-1}x^{(n-1)\frac{i}{k}} + a_{2n-2}x^{(n-2)\frac{i}{k}-g\sqrt{j}} + a_{2n-3}x^{(n-2)\frac{i}{k}} + ... + a_{3}x^{\frac{i}{k}} + a_{2}x^{-g\sqrt{j}} + a_{1} = 0$;

[9]: $a_{2n}x^{(n-1)(-g\sqrt{j})+\frac{i}{k}} + a_{2n-1}x^{(n-1)(-g\sqrt{j})} + a_{2n-2}x^{(n-2)(-g\sqrt{j})+\frac{i}{k}} + a_{2n-3}x^{(n-2)(-g\sqrt{j})} + ... + a_{3}x^{-g\sqrt{j}} + a_{2}x^{\frac{i}{k}} + a_{1} = 0$;

[10]: $a_{2n}x^{(n-1)(-\frac{i}{k})-\frac{j}{m}} + a_{2n-1}x^{(n-1)(-\frac{i}{k})} + a_{2n-2}x^{(n-2)(-\frac{i}{k})-\frac{j}{m}} + a_{2n-3}x^{(n-2)(-\frac{i}{k})} + ... + a_{3}x^{-\frac{i}{k}} + a_{2}x^{-\frac{j}{m}} + a_{1} = 0$;

[11]: $a_{2n}x^{(n-1)(-\frac{j}{m})+\frac{i}{k}} + a_{2n-1}x^{(n-1)(-\frac{j}{m})} + a_{2n-2}x^{(n-2)(-\frac{j}{m})+\frac{i}{k}} + a_{2n-3}x^{(n-2)(-\frac{j}{m})} + ... + a_{3}x^{-\frac{j}{m}} + a_{2}x^{\frac{i}{k}} + a_{1} = 0$;

[12]: $a_{2n}x^{(n-1)(-\frac{i}{k})-\frac{\sqrt{j}}{m}} + a_{2n-1}x^{(n-1)(-\frac{i}{k})} + a_{2n-2}x^{(n-2)(-\frac{i}{k})-\frac{\sqrt{j}}{m}} + a_{2n-3}x^{(n-2)(-\frac{i}{k})} + ... + a_{3}x^{-\frac{i}{k}} + a_{2}x^{-\frac{\sqrt{j}}{m}} + a_{1} = 0$;

[13]: $a_{2n}x^{(n-1)(-\frac{i}{k})-\frac{g\sqrt{j}}{m}} + a_{2n-1}x^{(n-1)(-\frac{i}{k})} + a_{2n-2}x^{(n-2)(-\frac{i}{k})-\frac{g\sqrt{j}}{m}} + a_{2n-3}x^{(n-2)(-\frac{i}{k})} + ... + a_{3}x^{-\frac{i}{k}} + a_{2}x^{-\frac{g\sqrt{j}}{m}} + a_{1} = 0$;

[14]: $a_{2n}x^{(n-1)(-\frac{\sqrt{j}}{m})-\frac{i}{k}} + a_{2n-1}x^{(n-1)(-\frac{\sqrt{j}}{m})} + a_{2n-2}x^{(n-2)(-\frac{\sqrt{j}}{m})-\frac{i}{k}} + a_{2n-3}x^{(n-2)(-\frac{\sqrt{j}}{m})} + ... + a_{3}x^{-\frac{\sqrt{j}}{m}} + a_{2}x^{-\frac{i}{k}} + a_{1} = 0$;

[15]: $a_{2n}x^{(n-1)(-\frac{g\sqrt{j}}{m})-\frac{i}{k}} + a_{2n-1}x^{(n-1)(-\frac{g\sqrt{j}}{m})} + a_{2n-2}x^{(n-2)(-\frac{g\sqrt{j}}{m})-\frac{i}{k}} + a_{2n-3}x^{(n-2)(-\frac{g\sqrt{j}}{m})} + ... + a_{3}x^{-\frac{g\sqrt{j}}{m}} + a_{2}x^{-\frac{i}{k}} + a_{1} = 0$;

[16]: $a_{2n}x^{(n-1)(-\frac{i}{k})-\sqrt{j}} + a_{2n-1}x^{(n-1)(-\frac{i}{k})} + a_{2n-2}x^{(n-2)(-\frac{i}{k})-\sqrt{j}} + a_{2n-3}x^{(n-2)(-\frac{i}{k})} + ... + a_{3}x^{-\frac{i}{k}} + a_{2}x^{-\sqrt{j}} + a_{1} = 0$;

[17]: $a_{2n}x^{(n-1)(-\frac{i}{k})-g\sqrt{j}} + a_{2n-1}x^{(n-1)(-\frac{i}{k})} + a_{2n-2}x^{(n-2)(-\frac{i}{k})-g\sqrt{j}} + a_{2n-3}x^{(n-2)(-\frac{i}{k})} + ... + a_{3}x^{-\frac{i}{k}} + a_{2}x^{-g\sqrt{j}} + a_{1} = 0$;

[18]: $a_{2n}x^{(n-1)(-\sqrt{j})-\frac{i}{k}} + a_{2n-1}x^{(n-1)(-\sqrt{j})} + a_{2n-2}x^{(n-2)(-\sqrt{j})-\frac{i}{k}} + a_{2n-3}x^{(n-2)(-\sqrt{j})} + ... + a_{3}x^{-\sqrt{j}} + a_{2}x^{-\frac{i}{k}} + a_{1} = 0$;

[19]: $a_{2n}x^{(n-1)(-g\sqrt{j})-\frac{i}{k}} + a_{2n-1}x^{(n-1)(-g\sqrt{j})} + a_{2n-2}x^{(n-2)(-g\sqrt{j})-\frac{i}{k}} + a_{2n-3}x^{(n-2)(-g\sqrt{j})} + ... + a_{3}x^{-g\sqrt{j}} + a_{2}x^{-\frac{i}{k}} + a_{1} = 0$;

[20]: $a_{2n}x^{(n-1)(-\frac{i}{k})-\frac{j}{m}} + a_{2n-1}x^{(n-1)(-\frac{i}{k})} + a_{2n-2}x^{(n-2)(-\frac{i}{k})-\frac{j}{m}} + a_{2n-3}x^{(n-2)(-\frac{i}{k})} + ... + a_{3}x^{-\frac{i}{k}} + a_{2}x^{-\frac{j}{m}} + a_{1} = 0$;

$$[21]: \; a_{2n}x^{(n-1)(-\frac{j}{m})-\frac{i}{k}} + a_{2n-1}x^{(n-1)(-\frac{j}{m})} + a_{2n-2}x^{(n-2)(-\frac{j}{m})-\frac{i}{k}} + a_{2n-3}x^{(n-2)(-\frac{j}{m})} + ... + a_3 x^{-\frac{j}{m}} + a_2 x^{-\frac{i}{k}} + a_1 = 0$$

$$(n = 2,3,4,...,\infty; i = 2,3,4,...,\infty, m = 2,3,4,...,\infty; j = 2,3,4,...,\infty; g = 2,3,4,...,\infty)$$

… …

About the Complex Composite Dinbakish Equations, I have no time to show them in the Volume 2. In fact, they have other things in which they are always great.

6: The Multivariate Dinbakish Equations

❶ : Ones of the general forms of the Multivariate Dinbakish Equations

i:
$$a_n (xy)^{(n)\frac{i}{k}} + a_{n-1}(xy)^{(n-1)\frac{i}{k}} + a_{n-2}(xy)^{(n-2)\frac{i}{k}} + ... + a_1 (xy)^{\frac{i}{k}} + a_0 = 0$$

ii:
$$a_n (xy)^{(n)(-\frac{i}{k})} + a_{n-1}(xy)^{(n-1)(-\frac{i}{k})} + a_{n-2}(xy)^{(n-2)(-\frac{i}{k})} + ... + a_1 (xy)^{(-\frac{i}{k})} + a_0 = 0 \; ;$$

iii:
$$a_n (xy)^{(n)\frac{\sqrt{i}}{k}} + a_{n-1}(xy)^{(n-1)\frac{\sqrt{i}}{k}} + a_{n-2}(xy)^{(n-2)\frac{\sqrt{i}}{k}} + ... + a_1 (xy)^{\frac{\sqrt{i}}{k}} + a_0 = 0 \; ;$$

iv:
$$a_n (xy)^{(n)(-\frac{\sqrt{i}}{k})} + a_{n-1}(xy)^{(n-1)(-\frac{\sqrt{i}}{k})} + a_{n-2}(xy)^{(n-2)(-\frac{\sqrt{i}}{k})} + ... + a_1 (xy)^{(-\frac{\sqrt{i}}{k})} + a_0 = 0 \; ;$$

v:
$$a_n (xy)^{(n)\sqrt{i}} + a_{n-1}(xy)^{(n-1)\sqrt{i}} + a_{n-2}(xy)^{(n-2)\sqrt{i}} + ... + a_1 (xy)^{\sqrt{i}} + a_0 = 0 \; ;$$

vi:
$$a_n (xy)^{(n)(-\sqrt{i})} + a_{n-1}(xy)^{(n-1)(-\sqrt{i})} + a_{n-2}(xy)^{(n-2)(-\sqrt{i})} + ... + a_1 (xy)^{(-\sqrt{i})} + a_0 = 0 \; ;$$

vii:
$$a_n (xy)^{(n)\frac{m\sqrt{i}}{k}} + a_{n-1}(xy)^{(n-1)\frac{m\sqrt{i}}{k}} + a_{n-2}(xy)^{(n-2)\frac{m\sqrt{i}}{k}} + ... + a_1 (xy)^{\frac{m\sqrt{i}}{k}} + a_0 = 0 \; ;$$

viii:
$$a_n (xy)^{(n)(-\frac{m\sqrt{i}}{k})} + a_{n-1}(xy)^{(n-1)(-\frac{m\sqrt{i}}{k})} + a_{n-2}(xy)^{(n-2)(-\frac{m\sqrt{i}}{k})} + ... + a_1 (xy)^{(-\frac{m\sqrt{i}}{k})} + a_0 = 0 \; ;$$

ix:
$$a_n (xy)^{(n)m\sqrt{i}} + a_{n-1}(xy)^{(n-1)m\sqrt{i}} + a_{n-2}(xy)^{(n-2)m\sqrt{i}} + ... + a_1 (xy)^{m\sqrt{i}} + a_0 = 0 \; ;$$

x:
$$a_n(xy)^{(n)(-m\sqrt{i})} + a_{n-1}(xy)^{(n-1)(-m\sqrt{i})} + a_{n-2}(xy)^{(n-2)(-m\sqrt{i})} + ... + a_1(xy)^{(-m\sqrt{i})} + a_0 = 0 \ ;$$

$$\left(y = a_1 x + a_0; i = 2,3,4,...,\infty; k = 2,3,4,...,\infty; m = 2,3,4,...,\infty; n = 2,3,4,...,\infty\right)$$

(1) : A Multivariate Dinbakish Equation

$$2(xy) - 3(xy)^{\frac{1}{2}} - 6 = 0 \ , \ y = -3x - 6 \ ;$$

This Multivariate Dinbakish Equation can be:

$$\left[(xy)^{\frac{1}{2}} - \frac{3}{4}\right]^2 = \frac{57}{16} \ ,$$

In which it has two parts:

Part 1: $xy = \left(\dfrac{\sqrt{57}+3}{4}\right)^2$; Part 2: $xy = \left(\dfrac{-\sqrt{57}+3}{4}\right)^2$.

In the part 1: $\quad y = \dfrac{\left(\dfrac{\sqrt{57}+3}{4}\right)^2}{x}$, $y = -3x - 6$;

It means that: $3x^2 + 6x + \left(\dfrac{\sqrt{57}+3}{4}\right)^2 = 0$; $(x+1)^2 = \dfrac{3 - \left(\dfrac{\sqrt{57}+3}{4}\right)^2}{3}$;

$$x_1 = -1 + \sqrt{\dfrac{3 - \left(\dfrac{\sqrt{57}+3}{4}\right)^2}{3}} \ , \ y_1 = -3 - 3\sqrt{\dfrac{3 - \left(\dfrac{\sqrt{57}+3}{4}\right)^2}{3}} \ ;$$

$$x_2 = -1 - \sqrt{\dfrac{3 - \left(\dfrac{\sqrt{57}+3}{4}\right)^2}{3}} \ , \ y_2 = -3 + 3\sqrt{\dfrac{3 - \left(\dfrac{\sqrt{57}+3}{4}\right)^2}{3}} \ .$$

42

In the part 2:

$$y = \dfrac{\left(\dfrac{-\sqrt{57}+3}{4}\right)^2}{x} \ , \ y = -3x - 6 \ ;$$

It means that:

$$3x^2 + 6x + \left(\dfrac{-\sqrt{57}+3}{4}\right)^2 = 0 \ , \ (x+1)^2 = \dfrac{3 - \left(\dfrac{-\sqrt{57}+3}{4}\right)^2}{3} \ ;$$

$$x_3 = -1 + \sqrt{\dfrac{3 - \left(\dfrac{-\sqrt{57}+3}{4}\right)^2}{3}} \ , \ y_3 = -3 - 3\sqrt{\dfrac{3 - \left(\dfrac{-\sqrt{57}+3}{4}\right)^2}{3}} \ ;$$

$$x_4 = -1 - \sqrt{\dfrac{3 - \left(\dfrac{-\sqrt{57}+3}{4}\right)^2}{3}} \ , \ y_4 = -3 + 3\sqrt{\dfrac{3 - \left(\dfrac{-\sqrt{57}+3}{4}\right)^2}{3}} \ .$$

(2) : A Multivariate Dinbakish Equation

$$(xy)^{-\frac{3}{2}} - 7(xy)^{-\frac{3}{4}} + 12 = 0 \ , \ y = -7x + 12 \ ;$$

This Multivariate Dinbakish Equation can be:

$$\left[(xy)^{-\frac{3}{4}} - 3\right]\left[(xy)^{-\frac{3}{4}} - 4\right] = 0 \ ,$$

In which it has two parts:

Part 1: $xy = 3^{-\frac{4}{3}}$; Part 2: $xy = 4^{-\frac{4}{3}}$.

In the part 1: $y = \dfrac{3^{-\frac{4}{3}}}{x}$, $y = -7x + 12$;

$$7x^2 - 12x + 3^{-\frac{4}{3}} = 0 \ , \ \left(x - \dfrac{6}{7}\right)^2 = \dfrac{36 - 7 \times 3^{-\frac{4}{3}}}{49} \ ;$$

$$x_1 = \dfrac{6 + \sqrt{36 - 7 \times 3^{-\frac{4}{3}}}}{7} \ , \ y_1 = 6 - \sqrt{36 - 7 \times 3^{-\frac{4}{3}}} \ ;$$

$$x_2 = \dfrac{6 - \sqrt{36 - 7 \times 3^{-\frac{4}{3}}}}{7} \ , \ y_2 = 6 + \sqrt{36 - 7 \times 3^{-\frac{4}{3}}} \ .$$

In the part 2:

$$y = \dfrac{4^{-\frac{4}{3}}}{x} \ , \ y = -7x + 12 \ ;$$

$$7x^2 - 12x + 4^{-\frac{4}{3}} = 0 \ , \ \left(x - \dfrac{6}{7}\right)^2 = \dfrac{36 - 7 \times 4^{-\frac{4}{3}}}{49} \ ;$$

$$x_3 = \dfrac{6 + \sqrt{36 - 7 \times 4^{-\frac{4}{3}}}}{7} \ , \ y_3 = 6 - \sqrt{36 - 7 \times 4^{-\frac{4}{3}}} \ ;$$

$$x_4 = \dfrac{6 - \sqrt{36 - 7 \times 4^{-\frac{4}{3}}}}{7} \ , \ y_4 = 6 + \sqrt{36 - 7 \times 4^{-\frac{4}{3}}} \ .$$

(3) : A Multivariate Dinbakish Equation

$$2(xy)^{\sqrt{5}} - \dfrac{4}{3}(xy)^{\frac{\sqrt{5}}{2}} + 7 = 0 \ , \ y = -\dfrac{4}{3}x + 7 \ ;$$

This Multivariate Dinbakish Equation can be:

$$\left[(xy)^{\frac{\sqrt{5}}{2}}-\frac{1}{3}\right]^2=-\frac{61}{18}\ ,$$

In which it has two parts:

$$\text{Part 1: } xy=\left(\frac{1+\sqrt{-\frac{61}{2}}}{3}\right)^{\frac{2\sqrt{5}}{5}}\quad;\quad \text{Part 2: } xy=\left(\frac{1-\sqrt{-\frac{61}{2}}}{3}\right)^{\frac{2\sqrt{5}}{5}}\ .$$

In the part 1:

$$y=\frac{\left(\dfrac{1+\sqrt{-\frac{61}{2}}}{3}\right)^{\frac{2\sqrt{5}}{5}}}{x}\ ,\ y=-\frac{4}{3}x+7\ ;$$

$$\frac{4}{3}x^2-7x+\left(\frac{1+\sqrt{-\frac{61}{2}}}{3}\right)^{\frac{2\sqrt{5}}{5}}=0\ ,\ \left(x-\frac{21}{8}\right)^2=\frac{441-48\left(\dfrac{1+\sqrt{-\frac{61}{2}}}{3}\right)^{\frac{2\sqrt{5}}{5}}}{64}\ ,$$

$$x_1=\frac{21+\sqrt{441-48\left(\dfrac{1+\sqrt{-\frac{61}{2}}}{3}\right)^{\frac{2\sqrt{5}}{5}}}}{8}\ ,\ y_1=\frac{21+\sqrt{441-48\left(\dfrac{1+\sqrt{-\frac{61}{2}}}{3}\right)^{\frac{2\sqrt{5}}{5}}}}{6}+7\ ;$$

$$x_2 = \frac{\sqrt{21 - \sqrt{441 - 48\left(\dfrac{1 + \sqrt{-\dfrac{61}{2}}}{3}\right)^{\frac{2\sqrt{5}}{5}}}}}{8} \ , \ y_2 = -\frac{\sqrt{21 - \sqrt{441 - 48\left(\dfrac{1 + \sqrt{-\dfrac{61}{2}}}{3}\right)^{\frac{2\sqrt{5}}{5}}}}}{6} + 7 \ .$$

In the part 2:

$$y = \frac{\left(\dfrac{1 - \sqrt{-\dfrac{61}{2}}}{3}\right)^{\frac{2\sqrt{5}}{5}}}{x} \ , \ y = -\frac{4}{3}x + 7 \ ;$$

$$\frac{4}{3}x^2 - 7x + \left(\frac{1 - \sqrt{-\dfrac{61}{2}}}{3}\right)^{\frac{2\sqrt{5}}{5}} = 0 \ , \ \left(x - \frac{21}{8}\right)^2 = \frac{441 - 48\left(\dfrac{1 - \sqrt{-\dfrac{61}{2}}}{3}\right)^{\frac{2\sqrt{5}}{5}}}{64} \ ;$$

$$x_3 = \frac{\sqrt{21 + \sqrt{441 - 48\left(\dfrac{1 - \sqrt{-\dfrac{61}{2}}}{3}\right)^{\frac{2\sqrt{5}}{5}}}}}{8} \ , \ y_3 = -\frac{\sqrt{21 + \sqrt{441 - 48\left(\dfrac{1 - \sqrt{-\dfrac{61}{2}}}{3}\right)^{\frac{2\sqrt{5}}{5}}}}}{6} + 7 \ ;$$

$$x_4 = \frac{\sqrt{21 - \sqrt{441 - 48\left(\dfrac{1 - \sqrt{-\dfrac{61}{2}}}{3}\right)^{\frac{2\sqrt{5}}{5}}}}}{8} \ , \ y_4 = -\frac{\sqrt{21 - \sqrt{441 - 48\left(\dfrac{1 - \sqrt{-\dfrac{61}{2}}}{3}\right)^{\frac{2\sqrt{5}}{5}}}}}{6} + 7 \ .$$

$$\left(4\right) : \text{ A Multivariate Dinbakish Equation}$$

$$3(xy)^{-\frac{2\sqrt{6}}{3}} + \frac{7}{5}(xy)^{-\frac{\sqrt{6}}{3}} - 4 = 0, \ y = \frac{7}{5}x - 4$$

This Multivariate Dinbakish Equation can be:

$$\left[(xy)^{-\frac{\sqrt{6}}{3}} + \frac{7}{30}\right]^2 = \frac{1249}{900} \ ,$$

In which it has two parts:

$$\text{Part 1: } xy = \left(\frac{-7+\sqrt{1249}}{30}\right)^{-\frac{\sqrt{6}}{2}} \ ; \text{ Part 2: } xy = \left(\frac{-7-\sqrt{1249}}{30}\right)^{-\frac{\sqrt{6}}{2}} \ .$$

In the part 1:

$$y = \frac{\left(\dfrac{-7+\sqrt{1249}}{30}\right)^{-\frac{\sqrt{6}}{2}}}{x} \ , \ y = \frac{7}{5}x - 4 \ ;$$

$$\frac{7}{5}x^2 - 4x - \left(\frac{-7+\sqrt{1249}}{30}\right)^{-\frac{\sqrt{6}}{2}} = 0 \ , \ \left(x - \frac{10}{7}\right)^2 = \frac{100 + 35\left(\dfrac{-7+\sqrt{1249}}{30}\right)^{-\frac{\sqrt{6}}{2}}}{49} \ ;$$

$$x_1 = \frac{10 + \sqrt{100 + 35\left(\dfrac{-7+\sqrt{1249}}{30}\right)^{-\frac{\sqrt{6}}{2}}}}{7} \ , \ y_1 = \frac{10 + \sqrt{100 + 35\left(\dfrac{-7+\sqrt{1249}}{30}\right)^{-\frac{\sqrt{6}}{2}}}}{5} - 4 \ ;$$

$$x_2 = \frac{10 - \sqrt{100 + 35\left(\dfrac{-7+\sqrt{1249}}{30}\right)^{-\frac{\sqrt{6}}{2}}}}{7} \;,\; y_2 = \frac{10 - \sqrt{100 + 35\left(\dfrac{-7+\sqrt{1249}}{30}\right)^{-\frac{\sqrt{6}}{2}}}}{5} - 4 \;.$$

In the part 2:

$$y = \frac{\left(\dfrac{-7-\sqrt{1249}}{30}\right)^{-\frac{\sqrt{6}}{2}}}{x} \;,\; y = \frac{7}{5}x - 4 \;;$$

$$\frac{7}{5}x^2 - 4x - \left(\frac{-7-\sqrt{1249}}{30}\right)^{-\frac{\sqrt{6}}{2}} = 0 \;,\; \left(x - \frac{10}{7}\right)^2 = \frac{100 + 35\left(\dfrac{-7-\sqrt{1249}}{30}\right)^{-\frac{\sqrt{6}}{2}}}{49} \;;$$

$$x_3 = \frac{10 + \sqrt{100 + 35\left(\dfrac{-7-\sqrt{1249}}{30}\right)^{-\frac{\sqrt{6}}{2}}}}{7} \;,\; y_3 = \frac{10 + \sqrt{100 + 35\left(\dfrac{-7-\sqrt{1249}}{30}\right)^{-\frac{\sqrt{6}}{2}}}}{5} - 4 \;;$$

$$x_4 = \frac{10 - \sqrt{100 + 35\left(\dfrac{-7-\sqrt{1249}}{30}\right)^{-\frac{\sqrt{6}}{2}}}}{7} \;,\; y_4 = \frac{10 - \sqrt{100 + 35\left(\dfrac{-7-\sqrt{1249}}{30}\right)^{-\frac{\sqrt{6}}{2}}}}{5} - 4 \;.$$

(5) : A Multivariate Dinbakish Equation

$$-\frac{3}{2}(xy)^{2\sqrt{3}} + 5(xy)^{\sqrt{3}} + 6 = 0 \;,\; y = 5x + 6 \;;$$

This Multivariate Dinbakish Equation can be:

$$\left[(xy)^{\sqrt{3}} - \frac{5}{3}\right]^2 = \frac{61}{9} \; ,$$

In which it has two parts:

$$\text{Part 1: } xy = \left(\frac{5+\sqrt{61}}{3}\right)^{-\frac{\sqrt{3}}{3}} \; ; \quad \text{Part 2: } xy = \left(\frac{5-\sqrt{61}}{3}\right)^{-\frac{\sqrt{3}}{3}} \; .$$

In the part 1:

$$y = \frac{\left(\dfrac{5+\sqrt{61}}{3}\right)^{-\frac{\sqrt{3}}{3}}}{x} \; , \; y = 5x+6 \; ,$$

$$5x^2 + 6x - \left(\frac{5+\sqrt{61}}{3}\right)^{-\frac{\sqrt{3}}{3}} = 0 \; , \; \left(x+\frac{3}{5}\right)^2 = \frac{9 + 5\left(\dfrac{5+\sqrt{61}}{3}\right)^{-\frac{\sqrt{3}}{3}}}{25} \; ;$$

$$x_1 = \frac{-3 + \sqrt{9 + 5\left(\dfrac{5+\sqrt{61}}{3}\right)^{-\frac{\sqrt{3}}{3}}}}{5} \; , \; y_1 = 3 + \sqrt{9 + 5\left(\dfrac{5+\sqrt{61}}{3}\right)^{-\frac{\sqrt{3}}{3}}} \; ;$$

$$x_2 = \frac{-3 - \sqrt{9 + 5\left(\dfrac{5+\sqrt{61}}{3}\right)^{-\frac{\sqrt{3}}{3}}}}{5} \; , \; y_2 = 3 - \sqrt{9 + 5\left(\dfrac{5+\sqrt{61}}{3}\right)^{-\frac{\sqrt{3}}{3}}} \; .$$

In the part 2:

$$xy = \left(\frac{5-\sqrt{61}}{3}\right)^{-\frac{\sqrt{3}}{3}} \; , \; y = 5x+6 \; ;$$

$$5x^2 + 6x - \left(\frac{5-\sqrt{61}}{3}\right)^{-\frac{\sqrt{3}}{3}} = 0 \ , \ \left(x+\frac{3}{5}\right)^2 = \frac{9+5\left(\frac{5-\sqrt{61}}{3}\right)^{-\frac{\sqrt{3}}{3}}}{25} \ ;$$

$$x_3 = \frac{-3+\sqrt{9+5\left(\frac{5-\sqrt{61}}{3}\right)^{-\frac{\sqrt{3}}{3}}}}{5} \ , \ y_3 = 3+\sqrt{9+5\left(\frac{5-\sqrt{61}}{3}\right)^{-\frac{\sqrt{3}}{3}}} \ ;$$

$$x_4 = \frac{-3-\sqrt{9+5\left(\frac{5-\sqrt{61}}{3}\right)^{-\frac{\sqrt{3}}{3}}}}{5} \ , \ y_4 = 3-\sqrt{9+5\left(\frac{5-\sqrt{61}}{3}\right)^{-\frac{\sqrt{3}}{3}}} \ .$$

(6) : A Multivariate Dinbakish Equation

$$(xy)^{-2\sqrt{7}} - \frac{1}{6}(xy)^{-\sqrt{7}} - \frac{5}{2} = 0 \ , \ y = -\frac{1}{6}x - \frac{5}{2} \ ;$$

This Multivariate Dinbakish Equation can be:

$$\left[(xy)^{-\sqrt{7}} + \frac{3}{2}\right]\left[(xy)^{-\sqrt{7}} - \frac{5}{3}\right] = 0 \ ,$$

In which it has two parts:

Part 1: $xy = \left(-\frac{3}{2}\right)^{-\frac{\sqrt{7}}{7}}$; Part 2: $xy = \left(\frac{5}{3}\right)^{-\frac{\sqrt{7}}{7}}$.

In the part 1:

$$xy = \left(-\frac{3}{2}\right)^{-\frac{\sqrt{7}}{7}} \ , \ y = -\frac{1}{6}x - \frac{5}{2} \ ;$$

$$x^2 + 15x + 6\left(-\frac{3}{2}\right)^{-\frac{\sqrt{7}}{7}} = 0 \ , \ \left(x + \frac{15}{2}\right)^2 = -\frac{225 + 24\left(-\frac{3}{2}\right)^{-\frac{\sqrt{7}}{7}}}{4} \ ;$$

$$x_1 = \frac{-15 + \sqrt{-225 - 24\left(-\frac{3}{2}\right)^{-\frac{\sqrt{7}}{7}}}}{2} \ , \ y_1 = \frac{-15 - \sqrt{-225 - 24\left(-\frac{3}{2}\right)^{-\frac{\sqrt{7}}{7}}}}{12} \ ;$$

$$x_2 = \frac{-15 - \sqrt{-225 - 24\left(-\frac{3}{2}\right)^{-\frac{\sqrt{7}}{7}}}}{2} \ , \ y_2 = \frac{-15 + \sqrt{-225 - 24\left(-\frac{3}{2}\right)^{-\frac{\sqrt{7}}{7}}}}{12} \ ;$$

In the part 2:

$$xy = \left(\frac{5}{3}\right)^{-\frac{\sqrt{7}}{7}} \ , \ y = -\frac{1}{6}x - \frac{5}{2} \ ;$$

$$x^2 + 15x + 6\left(\frac{5}{3}\right)^{-\frac{\sqrt{7}}{7}} = 0 \ , \ \left(x + \frac{15}{2}\right)^2 = -\frac{225 + 24\left(\frac{5}{3}\right)^{-\frac{\sqrt{7}}{7}}}{4} \ ;$$

$$x_3 = \frac{-15 + \sqrt{-225 - 24\left(\frac{5}{3}\right)^{-\frac{\sqrt{7}}{7}}}}{2} \ , \ y_3 = \frac{-15 - \sqrt{-225 - 24\left(\frac{5}{3}\right)^{-\frac{\sqrt{7}}{7}}}}{12} \ ;$$

$$x_4 = \frac{-15 - \sqrt{-225 - 24\left(\frac{5}{3}\right)^{-\frac{\sqrt{7}}{7}}}}{2} \ , \ y_4 = \frac{-15 + \sqrt{-225 - 24\left(\frac{5}{3}\right)^{-\frac{\sqrt{7}}{7}}}}{12} \ .$$

(7) : A Multivariate Dinbakish Equation

$$\frac{4}{3}(xy)^{\frac{4\sqrt{3}}{5}} + \frac{13}{3}(xy)^{\frac{2\sqrt{3}}{5}} - 35 = 0 \ , \ y = \frac{13}{3}x - 35 \ ;$$

This Multivariate Dinbakish Equation can be:

$$\left[\frac{4}{3}(xy)^{\frac{2\sqrt{3}}{5}} - 5\right]\left[(xy)^{\frac{2\sqrt{3}}{5}} + 7\right] = 0 \ ,$$

In which it has two parts:

Part 1: $xy = \left(\frac{15}{4}\right)^{\frac{5\sqrt{3}}{6}}$; Part 2: $xy = (-7)^{\frac{5\sqrt{3}}{6}}$.

In the part 1:

$$xy = \left(\frac{15}{4}\right)^{\frac{5\sqrt{3}}{6}} \ , \ y = \frac{13}{3}x - 35 \ ;$$

$$x^2 - \frac{105}{13}x - \frac{3}{13}\left(\frac{15}{4}\right)^{\frac{5\sqrt{3}}{6}} = 0 \ , \ \left(x - \frac{105}{26}\right)^2 = \frac{11025 + 156\left(\frac{15}{4}\right)^{\frac{5\sqrt{3}}{6}}}{676} \ ;$$

$$x_1 = \frac{105 + \sqrt{11025 + 156\left(\frac{15}{4}\right)^{\frac{5\sqrt{3}}{6}}}}{26} \ , \ y_1 = \frac{-105 + \sqrt{11025 + 156\left(\frac{15}{4}\right)^{\frac{5\sqrt{3}}{6}}}}{6} \ ;$$

$$x_2 = \frac{105 - \sqrt{11025 + 156\left(\frac{15}{4}\right)^{\frac{5\sqrt{3}}{6}}}}{26} \ , \ y_2 = \frac{-105 - \sqrt{11025 + 156\left(\frac{15}{4}\right)^{\frac{5\sqrt{3}}{6}}}}{6} \ .$$

In the part 2:

$$xy = (-7)^{\frac{5\sqrt{3}}{6}} \ , \ y = \frac{13}{3}x - 35 \ ;$$

$$x^2 - \frac{105}{13}x - \frac{3}{13}(-7)^{\frac{5\sqrt{3}}{6}} = 0 \ , \ \left(x - \frac{105}{26}\right)^2 = \frac{11025 + 156(-7)^{\frac{5\sqrt{3}}{6}}}{676} \ ;$$

$$x_3 = \frac{105 + \sqrt{11025 + 156(-7)^{\frac{5\sqrt{3}}{6}}}}{26} \ , \ y_3 = \frac{-105 + \sqrt{11025 + 156(-7)^{\frac{5\sqrt{3}}{6}}}}{6} \ ;$$

$$x_4 = \frac{105 - \sqrt{11025 + 156(-7)^{\frac{5\sqrt{3}}{6}}}}{26} \ , \ y_4 = \frac{-105 - \sqrt{11025 + 156(-7)^{\frac{5\sqrt{3}}{6}}}}{6} \ .$$

(8) : A Multivariate Dinbakish Equation

$$-3(xy)^{-\frac{5\sqrt{5}}{3}} - (xy)^{-\frac{5\sqrt{5}}{6}} + 44 = 0 \ , \ y = x + 44 \ ;$$

This Multivariate Dinbakish Equation can be:

$$\left[-3(xy)^{-\frac{5\sqrt{5}}{6}} + 11\right]\left[(xy)^{-\frac{5\sqrt{5}}{6}} + 4\right] = 0 \ ,$$

In which it has two parts:

Part 1: $xy = \left(\dfrac{11}{3}\right)^{-\frac{6\sqrt{5}}{25}}$; Part 2: $xy = (-4)^{-\frac{6\sqrt{5}}{25}}$.

In the part 1:

$$y = \frac{\left(\dfrac{11}{3}\right)^{-\frac{6\sqrt{5}}{25}}}{x} \ , \ y = x + 44 \ ;$$

$$x^2 + 44x - \left(\frac{11}{3}\right)^{-\frac{6\sqrt{5}}{25}} = 0 \ , \ (x + 22)^2 = 484 + \left(\frac{11}{3}\right)^{-\frac{6\sqrt{5}}{25}} \ ,$$

$$x_1 = -22 + \sqrt{484 + \left(\frac{11}{3}\right)^{-\frac{6\sqrt{5}}{25}}} \quad , \quad y_1 = 22 + \sqrt{484 + \left(\frac{11}{3}\right)^{-\frac{6\sqrt{5}}{25}}} \quad ;$$

$$x_2 = -22 - \sqrt{484 + \left(\frac{11}{3}\right)^{-\frac{6\sqrt{5}}{25}}} \quad , \quad y_2 = 22 - \sqrt{484 + \left(\frac{11}{3}\right)^{-\frac{6\sqrt{5}}{25}}} \quad .$$

In the part 2:

$$y = \frac{(-4)^{-\frac{6\sqrt{5}}{25}}}{x} \quad , \quad y = x + 44 \quad ;$$

$$x^2 + 44x - (-4)^{-\frac{6\sqrt{5}}{25}} = 0 \quad , \quad (x+22)^2 = 484 + (-4)^{-\frac{6\sqrt{5}}{25}} \quad ;$$

$$x_3 = -22 + \sqrt{484 + (-4)^{-\frac{6\sqrt{5}}{25}}} \quad , \quad y_3 = 22 + \sqrt{484 + (-4)^{-\frac{6\sqrt{5}}{25}}} \quad ;$$

$$x_4 = -22 - \sqrt{484 + (-4)^{-\frac{6\sqrt{5}}{25}}} \quad , \quad y_4 = 22 - \sqrt{484 + (-4)^{-\frac{6\sqrt{5}}{25}}} \quad .$$

❷: The other Multivariate Dinbakish Equation

i: $a_n (xyz)^{(n)\frac{i}{k}} + a_{n-1} (xyz)^{(n-1)\frac{i}{k}} + a_{n-2} (xyz)^{(n-2)\frac{i}{k}} + ... + a_1 (xyz)^{\frac{i}{k}} + a_0 = 0 \ ;$

ii: $a_n (xyz)^{(n)\left(-\frac{i}{k}\right)} + a_{n-1} (xyz)^{(n-1)(-\frac{i}{k})} + a_{n-2} (xyz)^{(n-2)(-\frac{i}{k})} + ... + a_1 (xyz)^{-\frac{i}{k}} + a_0 = 0 \ ;$

iii: $a_n (xyz)^{(n)\sqrt{i}} + a_{n-1} (xyz)^{(n-1)\sqrt{i}} + a_{n-2} (xyz)^{(n-2)\sqrt{i}} + ... + a_1 (xyz)^{\sqrt{i}} + a_0 = 0 \ ;$

iv: $a_n (xyz)^{(n)(-\sqrt{i})} + a_{n-1} (xyz)^{(n-1)(-\sqrt{i})} + a_{n-2} (xyz)^{(n-2)(-\sqrt{i})} + ... + a_1 (xyz)^{-\sqrt{i}} + a_0 = 0 \ ;$

v: $a_n (xyz)^{(n)\frac{\sqrt{i}}{k}} + a_{n-1} (xyz)^{(n-1)\frac{\sqrt{i}}{k}} + a_{n-2} (xyz)^{(n-2)\frac{\sqrt{i}}{k}} + ... + a_1 (xyz)^{\frac{\sqrt{i}}{k}} + a_0 = 0 \ ;$

vi: $a_n (xyz)^{(n)(-\frac{\sqrt{i}}{k})} + a_{n-1} (xyz)^{(n-1)(-\frac{\sqrt{i}}{k})} + a_{n-2} (xyz)^{(n-2)(-\frac{\sqrt{i}}{k})} + ... + a_1 (xyz)^{-\frac{\sqrt{i}}{k}} + a_0 = 0 \ ;$

$$(y = \sqrt{x} + a_1 \ , \ z = \sqrt{x} - a_1 \ , \ i = 2,3,4,...,\infty; k = 2,3,4,...,\infty; n = 2,3,4,...,\infty)$$

7: the Composite-Multivariate Dinbakish Equations

❶ : Ones of the general forms of the Composite-Multivariate Dinbakish Equations

$[1]$: $a_{2n}(xy)^{(n-1)\frac{i}{k}+\frac{\sqrt{j}}{k}} + a_{2n-1}(xy)^{(n-1)\frac{i}{k}} + a_{2n-2}(xy)^{(n-2)\frac{i}{k}+\frac{\sqrt{j}}{k}} + a_{2n-3}(xy)^{(n-2)\frac{i}{k}} + ... + a_3(xy)^{\frac{i}{k}} + a_2(xy)^{\frac{\sqrt{j}}{k}} + a_1 = 0$;

$[2]$: $a_{2n}(xy)^{(n-1)\frac{i}{k}+\frac{\sqrt{j}}{m}} + a_{2n-1}(xy)^{(n-1)\frac{i}{k}} + a_{2n-2}(xy)^{(n-2)\frac{i}{k}+\frac{\sqrt{j}}{m}} + a_{2n-3}(xy)^{(n-2)\frac{i}{k}} + ... + a_3(xy)^{\frac{i}{k}} + a_2(xy)^{\frac{\sqrt{j}}{m}} + a_1 = 0$;

$[3]$: $a_{2n}(xy)^{(n-1)\frac{i}{k}-\frac{\sqrt{j}}{k}} + a_{2n-1}(xy)^{(n-1)\frac{i}{k}} + a_{2n-2}(xy)^{(n-2)\frac{i}{k}-\frac{\sqrt{j}}{k}} + a_{2n-3}(xy)^{(n-2)\frac{i}{k}} + ... + a_3(xy)^{\frac{i}{k}} + a_2(xy)^{-\frac{\sqrt{j}}{k}} + a_1 = 0$;

$[4]$: $a_{2n}(xy)^{(n-1)\frac{i}{k}-\frac{\sqrt{j}}{m}} + a_{2n-1}(xy)^{(n-1)\frac{i}{k}} + a_{2n-2}(xy)^{(n-2)\frac{i}{k}-\frac{\sqrt{j}}{m}} + a_{2n-3}(xy)^{(n-2)\frac{i}{k}} + ... + a_3(xy)^{\frac{i}{k}} + a_2(xy)^{-\frac{\sqrt{j}}{m}} + a_1 = 0$;

$[5]$: $a_{2n}(xy)^{(n-1)(-\frac{i}{k})-\frac{\sqrt{j}}{k}} + a_{2n-1}(xy)^{(n-1)(-\frac{i}{k})} + a_{2n-2}(xy)^{(n-2)(-\frac{i}{k})-\frac{\sqrt{j}}{k}} + a_{2n-3}(xy)^{(n-2)(-\frac{i}{k})} + ... + a_3(xy)^{-\frac{i}{k}} + a_2(xy)^{-\frac{\sqrt{j}}{k}} + a_1 = 0$;

$[6]$: $a_{2n}(xy)^{(n-1)(-\frac{i}{k})-\frac{\sqrt{j}}{m}} + a_{2n-1}(xy)^{(n-1)(-\frac{i}{k})} + a_{2n-2}(xy)^{(n-2)(-\frac{i}{k})-\frac{\sqrt{j}}{m}} + a_{2n-3}(xy)^{(n-2)(-\frac{i}{k})} + ... + a_3(xy)^{-\frac{i}{k}} + a_2(xy)^{-\frac{\sqrt{j}}{m}} + a_1 = 0$;

$[7]$: $a_{2n}(xy)^{(n-1)(-\frac{i}{k})+\frac{\sqrt{j}}{k}} + a_{2n-1}(xy)^{(n-1)(-\frac{i}{k})} + a_{2n-2}(xy)^{(n-2)(-\frac{i}{k})+\frac{\sqrt{j}}{k}} + a_{2n-3}(xy)^{(n-2)(-\frac{i}{k})} + ... + a_3(xy)^{-\frac{i}{k}} + a_2(xy)^{\frac{\sqrt{j}}{k}} + a_1 = 0$;

$[8]$: $a_{2n}(xy)^{(n-1)(-\frac{i}{k})+\frac{\sqrt{j}}{m}} + a_{2n-1}(xy)^{(n-1)(-\frac{i}{k})} + a_{2n-2}(xy)^{(n-2)(-\frac{i}{k})+\frac{\sqrt{j}}{m}} + a_{2n-3}(xy)^{(n-2)(-\frac{i}{k})} + ... + a_3(xy)^{-\frac{i}{k}} + a_2(xy)^{\frac{\sqrt{j}}{m}} + a_1 = 0$;

$[9]$: $a_{2n}(xy)^{(n-1)\frac{i}{k}+\sqrt{j}} + a_{2n-1}(xy)^{(n-1)\frac{i}{k}} + a_{2n-2}(xy)^{(n-2)\frac{i}{k}+\sqrt{j}} + a_{2n-3}(xy)^{(n-2)\frac{i}{k}} + ... + a_3(xy)^{\frac{i}{k}} + a_2(xy)^{\sqrt{j}} + a_1 = 0$;

$[10]$: $a_{2n}(xy)^{(n-1)\frac{i}{k}-\sqrt{j}} + a_{2n-1}(xy)^{(n-1)\frac{i}{k}} + a_{2n-2}(xy)^{(n-2)\frac{i}{k}-\sqrt{j}} + a_{2n-3}(xy)^{(n-2)\frac{i}{k}} + ... + a_3(xy)^{\frac{i}{k}} + a_2(xy)^{-\sqrt{j}} + a_1 = 0$;

$[11]$: $a_{2n}(xy)^{(n-1)(-\frac{i}{k})+\sqrt{j}} + a_{2n-1}(xy)^{(n-1)(-\frac{i}{k})} + a_{2n-2}(xy)^{(n-2)(-\frac{i}{k})+\sqrt{j}} + a_{2n-3}(xy)^{(n-2)(-\frac{i}{k})} + ... + a_3(xy)^{-\frac{i}{k}} + a_2(xy)^{\sqrt{j}} + a_1 = 0$;

$[12]$: $a_{2n}(xy)^{(n-1)(-\frac{i}{k})-\sqrt{j}} + a_{2n-1}(xy)^{(n-1)(-\frac{i}{k})} + a_{2n-2}(xy)^{(n-2)(-\frac{i}{k})-\sqrt{j}} + a_{2n-3}(xy)^{(n-2)(-\frac{i}{k})} + ... + a_3(xy)^{-\frac{i}{k}} + a_2(xy)^{-\sqrt{j}} + a_1 = 0$;

$[\,13\,]$: $a_{2n}(xy)^{(n-1)\sqrt{i}+\frac{\sqrt{j}}{k}}+a_{2n-1}(xy)^{(n-1)\sqrt{i}}+a_{2n-2}(xy)^{(n-2)\sqrt{i}+\frac{\sqrt{j}}{k}}+a_{2n-3}(xy)^{(n-2)\sqrt{i}}+...+a_{3}(xy)^{\sqrt{i}}+a_{2}(xy)^{\frac{\sqrt{j}}{k}}+a_{1}=0$;

$[14]$: $a_{2n}(xy)^{(n-1)\sqrt{i}-\frac{\sqrt{j}}{k}}+a_{2n-1}(xy)^{(n-1)\sqrt{i}}+a_{2n-2}(xy)^{(n-2)\sqrt{i}-\frac{\sqrt{j}}{k}}+a_{2n-3}(xy)^{(n-2)\sqrt{i}}+...+a_{3}(xy)^{\sqrt{i}}+a_{2}(xy)^{-\frac{\sqrt{j}}{k}}+a_{1}=0$;

$[15]$: $a_{2n}(xy)^{(n-1)(-\sqrt{i})+\frac{\sqrt{j}}{k}}+a_{2n-1}(xy)^{(n-1)(-\sqrt{i})}+a_{2n-2}(xy)^{(n-2)(-\sqrt{i})+\frac{\sqrt{j}}{k}}+a_{2n-3}(xy)^{(n-2)(-\sqrt{i})}+...+a_{3}(xy)^{-\sqrt{i}}+a_{2}(xy)^{\frac{\sqrt{j}}{k}}+a_{1}=0$;

$[16]$: $a_{2n}(xy)^{(n-1)(-\sqrt{i})-\frac{\sqrt{j}}{k}}+a_{2n-1}(xy)^{(n-1)(-\sqrt{i})}+a_{2n-2}(xy)^{(n-2)(-\sqrt{i})-\frac{\sqrt{j}}{k}}+a_{2n-3}(xy)^{(n-2)(-\sqrt{i})}+...+a_{3}(xy)^{-\sqrt{i}}+a_{2}(xy)^{-\frac{\sqrt{j}}{k}}+a_{1}=0$;

$[17]$: $a_{2n}(xy)^{(n-1)\sqrt{i}+\sqrt{j}}+a_{2n-1}(xy)^{(n-1)\sqrt{i}}+a_{2n-2}(xy)^{(n-2)\sqrt{i}+\sqrt{j}}+a_{2n-3}(xy)^{(n-2)\sqrt{i}}+...+a_{3}(xy)^{\sqrt{i}}+a_{2}(xy)^{\sqrt{j}}+a_{1}=0$;

$[18]$: $a_{2n}(xy)^{(n-1)\sqrt{i}-\sqrt{j}}+a_{2n-1}(xy)^{(n-1)\sqrt{i}}+a_{2n-2}(xy)^{(n-2)\sqrt{i}-\sqrt{j}}+a_{2n-3}(xy)^{(n-2)\sqrt{i}}+...+a_{3}(xy)^{\sqrt{i}}+a_{2}(xy)^{-\sqrt{j}}+a_{1}=0$;

$[19]$: $a_{2n}(xy)^{(n-1)(-\sqrt{i})+\sqrt{j}}+a_{2n-1}(xy)^{(n-1)(-\sqrt{i})}+a_{2n-2}(xy)^{(n-2)(-\sqrt{i})+\sqrt{j}}+a_{2n-3}(xy)^{(n-2)(-\sqrt{i})}+...+a_{3}(xy)^{-\sqrt{i}}+a_{2}(xy)^{\sqrt{j}}+a_{1}=0$;

$[20]$: $a_{2n}(xy)^{(n-1)(-\sqrt{i})-\sqrt{j}}+a_{2n-1}(xy)^{(n-1)(-\sqrt{i})}+a_{2n-2}(xy)^{(n-2)(-\sqrt{i})-\sqrt{j}}+a_{2n-3}(xy)^{(n-2)(-\sqrt{i})}+...+a_{3}(xy)^{-\sqrt{i}}+a_{2}(xy)^{-\sqrt{j}}+a_{1}=0$.

$$\left(i=2,3,4,...,\infty; n=2,3,4,...,\infty; j=2,3,4,...,\infty; m=2,3,4,...,\infty; k=2,3,4,...,\infty\right)$$

(1) : A Composite-Multivariate Dinbakish Equation

$$a_{6}(xy)^{\frac{2}{5}+\frac{\sqrt{3}}{5}}+a_{5}(xy)^{\frac{2}{5}}+a_{4}(xy)^{\frac{1}{5}+\frac{\sqrt{3}}{5}}+a_{3}(xy)^{\frac{1}{5}}+a_{2}(xy)^{\frac{\sqrt{3}}{5}}+a_{1}=0, \; y=a_{2}x+a_{1} \; ;$$

One of the forms of this Composite-Multivariate Dinbakish Equation can be:

$$\left[a_{6}(xy)^{\frac{1}{5}}+3\right]\left[(xy)^{\frac{1}{5}}+\frac{5}{a_{6}}\right]\left[(xy)^{\frac{\sqrt{3}}{5}}-5\right]=0,$$

It expands:

$$a_{6}(xy)^{\frac{2}{5}+\frac{\sqrt{3}}{5}}-5a_{6}(xy)^{\frac{2}{5}}+8(xy)^{\frac{1}{5}+\frac{\sqrt{3}}{5}}-40(xy)^{\frac{1}{5}}+\frac{15}{a_{6}}(xy)^{\frac{\sqrt{3}}{5}}-\frac{75}{a_{6}}=0,$$

And $a_{5}=-5a_{6}; a_{4}=8; a_{3}=-40; a_{2}=\dfrac{15}{a_{6}}; a_{1}=-\dfrac{75}{a_{6}}$.

The Composite-Multivariate Dinbakish Equation follows the real number a_6 to be changed.

When $a_6 = -5$, it means that:

$$a_5 = 25; a_2 = -3; a_1 = 15 .$$

Now, the Composite-Multivariate Dinbakish Equation is:

$$-5(xy)^{\frac{2}{5}+\frac{\sqrt{3}}{5}} + 25(xy)^{\frac{2}{5}} + 8(xy)^{\frac{1}{5}+\frac{\sqrt{3}}{5}} - 40(xy)^{\frac{1}{5}} - 3(xy)^{\frac{\sqrt{3}}{5}} + 15 = 0 , y = -3x + 15 .$$

In which it has three parts:

Part 1: $xy = \left(\dfrac{3}{5}\right)^5$; Part 2: $xy = 1$; Part 3: $xy = 5^{\frac{5\sqrt{3}}{3}}$.

In the part 1:

$$xy = \left(\frac{3}{5}\right)^5 , y = -3x + 15 ;$$

$$y = \frac{\left(\dfrac{3}{5}\right)^5}{x} , \left(x - \frac{5}{2}\right)^2 = \frac{75 - 4\left(\dfrac{3}{5}\right)^5}{12} ;$$

$$x_1 = \frac{5}{2} + \sqrt{\frac{75 - 4\left(\dfrac{3}{5}\right)^5}{12}} , y_1 = \frac{15}{2} - 3\sqrt{\frac{75 - 4\left(\dfrac{3}{5}\right)^5}{12}} ;$$

$$x_2 = \frac{5}{2} - \sqrt{\frac{75 - 4\left(\dfrac{3}{5}\right)^5}{12}} , y_2 = \frac{15}{2} + 3\sqrt{\frac{75 - 4\left(\dfrac{3}{5}\right)^5}{12}} .$$

In the part 2:

$$xy = 1 , y = -3x + 15 ;$$

57

$$y = \frac{1}{x} \ , \ \left(x - \frac{5}{2}\right)^2 = \frac{71}{12} \ ;$$

$$x_3 = \frac{5}{2} + \sqrt{\frac{71}{12}} \ , \ y_3 = \frac{15}{2} - 3\sqrt{\frac{71}{12}} \ ;$$

$$x_4 = \frac{5}{2} - \sqrt{\frac{71}{12}} \ , \ y_4 = \frac{15}{2} + 3\sqrt{\frac{71}{12}} \ .$$

In the part 3:

$$xy = 5^{\frac{5\sqrt{3}}{3}} \ , \ y = -3x + 15 \ ;$$

$$y = \frac{5^{\frac{5\sqrt{3}}{3}}}{x} \ , \ \left(x - \frac{5}{2}\right)^2 = \frac{25}{4} - \frac{5^{\frac{5\sqrt{3}}{3}}}{3} \ ;$$

$$x_5 = \frac{5}{2} + \sqrt{\frac{25}{4} - \frac{5^{\frac{5\sqrt{3}}{3}}}{3}} \ , \ y_5 = \frac{15}{2} - 3\sqrt{\frac{25}{4} - \frac{5^{\frac{5\sqrt{3}}{3}}}{3}} \ ;$$

$$x_6 = \frac{5}{2} - \sqrt{\frac{25}{4} - \frac{5^{\frac{5\sqrt{3}}{3}}}{3}} \ , \ y_6 = \frac{15}{2} + 3\sqrt{\frac{25}{4} - \frac{5^{\frac{5\sqrt{3}}{3}}}{3}} \ .$$

(2) : A Composite-Multivariate Dinbakish Equation

$$a_6 (xy)^{\frac{12}{5} - \sqrt{5}} + a_5 (xy)^{\frac{12}{5}} + a_4 (xy)^{\frac{6}{5} - \sqrt{5}} + a_3 (xy)^{\frac{6}{5}} + a_2 (xy)^{-\sqrt{5}} + a_1 = 0 \ , \ y = a_2 x + a_1 \ ;$$

One of the forms of this Composite-Multivariate Dinbakish Equation can be:

$$\left[a_6 (xy)^{\frac{12}{5}} + 2a_6 (xy)^{\frac{6}{5}} - 3 \right]\left[(xy)^{-\sqrt{5}} - \frac{6}{a_6} \right] = 0 \ ,$$

It expands:

$$a_6(xy)^{\frac{12}{5}-\sqrt{5}} - 6(xy)^{\frac{12}{5}} + 2a_6(xy)^{\frac{6}{5}-\sqrt{5}} - 12(xy)^{\frac{6}{5}} - 3(xy)^{-\sqrt{5}} + \frac{18}{a_6} = 0 \ ,$$

And $a_5 = -6; a_4 = 2a_6; a_3 = -12; a_2 = -3; a_1 = \dfrac{18}{a_6}$.

The Composite-Multivariate Dinbakish Equation follows the real number a_6 to be changed.

When $a_6 = -3$, it means that:

$$a_4 = -6; a_1 = -6 \ .$$

Now, the Composite-Multivariate Dinbakish Equation is:

$$-3(xy)^{\frac{12}{5}-\sqrt{5}} - 6(xy)^{\frac{12}{5}} - 6(xy)^{\frac{6}{5}-\sqrt{5}} - 12(xy)^{\frac{6}{5}} - 3(xy)^{-\sqrt{5}} - 6 = 0 \ , \ y = -3x - 6 \ ;$$

In which it has three parts:

Part 1: $xy = \left(-1 + \dfrac{\sqrt{2}}{2}\right)^{\frac{5}{6}}$;   Part 2: $xy = \left(-1 - \dfrac{\sqrt{2}}{2}\right)^{\frac{5}{6}}$; Part 3: $xy = (-1)^{-\frac{\sqrt{5}}{5}}$.

In the part 1:

$$xy = \left(-1 + \dfrac{\sqrt{2}}{2}\right)^{\frac{5}{6}} \ , \ y = -3x - 6 \ ;$$

$$y = \frac{\left(-1 + \dfrac{\sqrt{2}}{2}\right)^{\frac{5}{6}}}{x} \ , \ (x+1)^2 = \frac{3 - \left(-1 + \dfrac{\sqrt{2}}{2}\right)^{\frac{5}{6}}}{3} \ ;$$

$$x_1 = -1 + \sqrt{\frac{3 - \left(-1 + \dfrac{\sqrt{2}}{2}\right)^{\frac{5}{6}}}{3}} \ , \ y_1 = -3 - 3\sqrt{\frac{3 - \left(-1 + \dfrac{\sqrt{2}}{2}\right)^{\frac{5}{6}}}{3}} \ ;$$

$$x_2 = -1 - \sqrt{\dfrac{3 - \left(-1+\dfrac{\sqrt{2}}{2}\right)^{\frac{5}{6}}}{3}} \quad, \quad y_2 = -3 + 3\sqrt{\dfrac{3 - \left(-1+\dfrac{\sqrt{2}}{2}\right)^{\frac{5}{6}}}{3}} \quad.$$

In the part 2:

$$xy = \left(-1 - \dfrac{\sqrt{2}}{2}\right)^{\frac{5}{6}} \quad, \quad y = -3x - 6 \ ;$$

$$y = \dfrac{\left(-1 - \dfrac{\sqrt{2}}{2}\right)^{\frac{5}{6}}}{x} \quad, \quad (x+1)^2 = \dfrac{3 - \left(-1 - \dfrac{\sqrt{2}}{2}\right)^{\frac{5}{6}}}{3} \ ;$$

$$x_3 = -1 + \sqrt{\dfrac{3 - \left(-1-\dfrac{\sqrt{2}}{2}\right)^{\frac{5}{6}}}{3}} \quad, \quad y_3 = -3 - 3\sqrt{\dfrac{3 - \left(-1-\dfrac{\sqrt{2}}{2}\right)^{\frac{5}{6}}}{3}} \ ;$$

$$x_4 = -1 - \sqrt{\dfrac{3 - \left(-1-\dfrac{\sqrt{2}}{2}\right)^{\frac{5}{6}}}{3}} \quad, \quad y_4 = -3 + 3\sqrt{\dfrac{3 - \left(-1-\dfrac{\sqrt{2}}{2}\right)^{\frac{5}{6}}}{3}} \quad.$$

In the part 3:

$$xy = (-1)^{-\frac{\sqrt{5}}{5}} \quad, \quad y = -3x - 6 \ ;$$

$$y = \dfrac{(-1)^{-\frac{\sqrt{5}}{5}}}{x} \quad, \quad (x+1)^2 = \dfrac{3 - (-1)^{-\frac{\sqrt{5}}{5}}}{3} \ ;$$

$$x_5 = -1 + \sqrt{\dfrac{3 - (-1)^{-\frac{\sqrt{5}}{5}}}{3}} \quad, \quad y_5 = -3 - 3\sqrt{\dfrac{3 - (-1)^{-\frac{\sqrt{5}}{5}}}{3}} \ ;$$

$$x_6 = -1 - \sqrt{\dfrac{3-(-1)^{-\frac{\sqrt{5}}{5}}}{3}} \quad , \quad y_6 = -3 + 3\sqrt{\dfrac{3-(-1)^{-\frac{\sqrt{5}}{5}}}{3}} \quad .$$

(3) : A Composite-Multivariate Dinbakish Equation

$$a_6 (xy)^{-\frac{2\sqrt{7}}{3}+\frac{3}{4}} + a_5 (xy)^{-\frac{2\sqrt{7}}{3}} + a_4 (xy)^{-\frac{\sqrt{7}}{3}+\frac{3}{4}} + a_3 (xy)^{-\frac{\sqrt{7}}{3}} + a_2 (xy)^{\frac{3}{4}} + a_1 = 0 \ , \ y = a_2 x + a_1 \ ;$$

One of the forms of this Composite-Multivariate Dinbakish Equation can be:

$$\left[a_6 (xy)^{\frac{-2\sqrt{7}}{3}} + \frac{1}{3a_6} (xy)^{-\frac{\sqrt{7}}{3}} + 5 \right]\left[(xy)^{\frac{3}{4}} + 2a_6 \right] = 0 \ ,$$

It expands:

$$a_6 (xy)^{-\frac{2\sqrt{7}}{3}+\frac{3}{4}} + 2a_6^2 (xy)^{-\frac{2\sqrt{7}}{3}} + \frac{1}{3a_6} (xy)^{-\frac{\sqrt{7}}{3}+\frac{3}{4}} + \frac{2}{3} (xy)^{-\frac{\sqrt{7}}{3}} + 5 (xy)^{\frac{3}{4}} + 10a_6 = 0,$$

And $a_5 = 2a_6^2 ; a_4 = \dfrac{1}{3a_6} ; a_3 = \dfrac{2}{3} ; a_2 = 5 ; a_1 = 10a_6$.

The Composite-Multivariate Dinbakish Equation follows the real number a_6 to be changed.

When $a_6 = 2$, it means that:

$$a_5 = 8 ; a_4 = \frac{1}{6} ; a_1 = 20 \ .$$

Now, the Composite-Multivariate Dinbakish Equation is:

$$2(xy)^{-\frac{2\sqrt{7}}{3}+\frac{3}{4}} + 8(xy)^{-\frac{2\sqrt{7}}{3}} + \frac{1}{6}(xy)^{-\frac{\sqrt{7}}{3}+\frac{3}{4}} + \frac{2}{3}(xy)^{-\frac{\sqrt{7}}{3}} + 5(xy)^{\frac{3}{4}} + 20 = 0 \ , \ y = 5x + 20 \ .$$

In which it has three parts:

Part 1: $xy = \left(\dfrac{-1+\sqrt{-1439}}{24} \right)^{-\frac{3\sqrt{7}}{7}}$; Part 2: $xy = \left(\dfrac{-1-\sqrt{-1439}}{24} \right)^{-\frac{3\sqrt{7}}{7}}$;

Part 3: $xy = (-4)^{\frac{4}{3}}$.

In the part 1:

$$xy = \left(\frac{-1+\sqrt{-1439}}{24}\right)^{-\frac{3\sqrt{7}}{7}} \ , \ y = 5x + 20 \ ;$$

$$y = \frac{\left(\frac{-1+\sqrt{-1439}}{24}\right)^{-\frac{3\sqrt{7}}{7}}}{x} \ , \ (x+2)^2 = \frac{20 + \left(\frac{-1+\sqrt{-1439}}{24}\right)^{-\frac{3\sqrt{7}}{7}}}{5} \ ;$$

$$x_1 = -2 + \sqrt{\frac{20 + \left(\frac{-1+\sqrt{-1439}}{24}\right)^{-\frac{3\sqrt{7}}{7}}}{5}} \ , \ y_1 = 10 + 5\sqrt{\frac{20 + \left(\frac{-1+\sqrt{-1439}}{24}\right)^{-\frac{3\sqrt{7}}{7}}}{5}} \ ;$$

$$x_2 = -2 - \sqrt{\frac{20 + \left(\frac{-1+\sqrt{-1439}}{24}\right)^{-\frac{3\sqrt{7}}{7}}}{5}} \ , \ y_2 = 10 - 5\sqrt{\frac{20 + \left(\frac{-1+\sqrt{-1439}}{24}\right)^{-\frac{3\sqrt{7}}{7}}}{5}} \ .$$

In the part 2:

$$xy = \left(\frac{-1-\sqrt{-1439}}{24}\right)^{-\frac{3\sqrt{7}}{7}} \ , \ y = 5x + 20 \ ;$$

$$y = \frac{\left(\frac{-1-\sqrt{-1439}}{24}\right)^{-\frac{3\sqrt{7}}{7}}}{x} \ , \ (x+2)^2 = \frac{20 + \left(\frac{-1-\sqrt{-1439}}{24}\right)^{-\frac{3\sqrt{7}}{7}}}{5} \ ;$$

$$x_3 = -2 + \sqrt{\dfrac{20 + \left(\dfrac{-1-\sqrt{-1439}}{24}\right)^{-\frac{3\sqrt{7}}{7}}}{5}} \;,\; y_3 = 10 + 5\sqrt{\dfrac{20 + \left(\dfrac{-1-\sqrt{-1439}}{24}\right)^{-\frac{3\sqrt{7}}{7}}}{5}} \;;$$

$$x_4 = -2 - \sqrt{\dfrac{20 + \left(\dfrac{-1-\sqrt{-1439}}{24}\right)^{-\frac{3\sqrt{7}}{7}}}{5}} \;,\; y_4 = 10 - 5\sqrt{\dfrac{20 + \left(\dfrac{-1-\sqrt{-1439}}{24}\right)^{-\frac{3\sqrt{7}}{7}}}{5}} \;.$$

In the part 3:

$$xy = (-4)^{\frac{4}{3}} \;,\; y = 5x + 20 \;;$$

$$y = \dfrac{(-4)^{\frac{4}{3}}}{x} \;,\; (x+2)^2 = \dfrac{20 + (-4)^{\frac{4}{3}}}{5} \;;$$

$$x_5 = -2 + \sqrt{\dfrac{20 + (-4)^{\frac{4}{3}}}{5}} \;,\; y_5 = 10 + 5\sqrt{\dfrac{20 + (-4)^{\frac{4}{3}}}{5}} \;;$$

$$x_6 = -2 - \sqrt{\dfrac{20 + (-4)^{\frac{4}{3}}}{5}} \;,\; y_6 = 10 - 5\sqrt{\dfrac{20 + (-4)^{\frac{4}{3}}}{5}} \;.$$

(4) : A Composite-Multivariate Dinbakish Equation

$$a_6(xy)^{-\frac{2\sqrt{6}}{3}-\sqrt{2}} + a_5(xy)^{-\frac{2\sqrt{6}}{3}} + a_4(xy)^{-\frac{\sqrt{6}}{3}-\sqrt{2}} + a_3(xy)^{-\frac{\sqrt{6}}{3}} + a_2(xy)^{-\sqrt{2}} + a_1 = 0 \;,\; y = a_2 x + a_1 \;;$$

One of the forms of this Composite-Multivariate Dinbakish Equation can be:

$$\left[a_6(xy)^{-\frac{2\sqrt{6}}{3}} + 2a_6(xy)^{-\frac{\sqrt{6}}{3}} - 4 \right]\left[(xy)^{-\sqrt{2}} - \dfrac{a_6}{2} \right] = 0 \;,$$

It expands:

$$(xy)^{-\frac{2\sqrt{6}}{3}-\sqrt{2}} - \frac{a_6}{2}(xy)^{-\frac{2\sqrt{6}}{3}} + 2a_6(xy)^{-\frac{\sqrt{6}}{3}-\sqrt{2}} - a_6^2(xy)^{-\frac{\sqrt{6}}{3}} - 4(xy)^{-\sqrt{2}} + 2a_6 = 0 \ ,$$

And $a_5 = -\dfrac{a_6}{2}; a_4 = 2a_6; a_3 = -a_6^2; a_2 = 4; a_1 = 2a_6$.

The Composite-Multivariate Dinbakish Equation follows the real number a_6 to be changed.

Because $a_6 = 1$, it means that:

$$a_5 = -\frac{1}{2}; a_4 = 2; a_3 = -1; a_1 = 2 \ .$$

Now, the Composite-Multivariate Dinbakish Equation is:

$$(xy)^{-\frac{2\sqrt{6}}{3}-\sqrt{2}} - \frac{1}{2}(xy)^{-\frac{2\sqrt{6}}{3}} + 2(xy)^{-\frac{\sqrt{6}}{3}-\sqrt{2}} - (xy)^{-\frac{\sqrt{6}}{3}} - 4(xy)^{-\sqrt{2}} + 2 = 0 \ , \ y = -4x + 2 \ ;$$

In which it has three parts:

Part 1: $xy = \left(-1+\sqrt{5}\right)^{-\frac{\sqrt{6}}{2}}$; Part 2: $xy = \left(-1-\sqrt{5}\right)^{-\frac{\sqrt{6}}{2}}$; Part 3: $xy = \left(\dfrac{1}{2}\right)^{-\frac{\sqrt{2}}{2}}$.

In the part 1:

$$xy = \left(-1+\sqrt{5}\right)^{-\frac{\sqrt{6}}{2}} \ , \ y = -4x + 2 \ ;$$

$$y = \frac{\left(-1+\sqrt{5}\right)^{-\frac{\sqrt{6}}{2}}}{x} \ , \ \left(x+\frac{1}{4}\right)^2 = \frac{1+4\left(-1+\sqrt{5}\right)^{-\frac{\sqrt{6}}{2}}}{16} \ ;$$

$$x_1 = \frac{-1+\sqrt{1+4\left(-1+\sqrt{5}\right)^{-\frac{\sqrt{6}}{2}}}}{4} \ , \ y_1 = 1 - \sqrt{1+4\left(-1+\sqrt{5}\right)^{-\frac{\sqrt{6}}{2}}} \ ;$$

64

$$x_2 = \frac{-1-\sqrt{1+4\left(-1+\sqrt{5}\right)^{-\frac{\sqrt{6}}{2}}}}{4} \ , \ y_2 = 1+\sqrt{1+4\left(-1+\sqrt{5}\right)^{-\frac{\sqrt{6}}{2}}} \ .$$

In the part 2:

$$xy = \left(-1-\sqrt{5}\right)^{-\frac{\sqrt{6}}{2}} \ , \ y = -4x+2 \ ;$$

$$y = \frac{\left(-1-\sqrt{5}\right)^{-\frac{\sqrt{6}}{2}}}{x} \ , \ \left(x+\frac{1}{4}\right)^2 = \frac{1+4\left(-1-\sqrt{5}\right)^{-\frac{\sqrt{6}}{2}}}{16} \ ;$$

$$x_3 = \frac{-1+\sqrt{1+4\left(-1-\sqrt{5}\right)^{-\frac{\sqrt{6}}{2}}}}{4} \ , \ y_3 = 1-\sqrt{1+4\left(-1-\sqrt{5}\right)^{-\frac{\sqrt{6}}{2}}} \ ;$$

$$x_4 = \frac{-1-\sqrt{1+4\left(-1-\sqrt{5}\right)^{-\frac{\sqrt{6}}{2}}}}{4} \ , \ y_4 = 1+\sqrt{1+4\left(-1-\sqrt{5}\right)^{-\frac{\sqrt{6}}{2}}} \ .$$

In the part 3:

$$xy = \left(\frac{1}{2}\right)^{-\frac{\sqrt{2}}{2}} \ , \ y = -4x+2 \ ;$$

$$y = \frac{\left(\frac{1}{2}\right)^{-\frac{\sqrt{2}}{2}}}{x} \ , \ \left(x+\frac{1}{4}\right)^2 = \frac{1+4\left(\frac{1}{2}\right)^{-\frac{\sqrt{2}}{2}}}{16} \ ;$$

$$x_5 = \frac{-1+\sqrt{1+4\left(\frac{1}{2}\right)^{-\frac{\sqrt{2}}{2}}}}{4} \ , \ y_5 = 1-\sqrt{1+4\left(\frac{1}{2}\right)^{-\frac{\sqrt{2}}{2}}} \ ;$$

$$x_6 = \dfrac{-1 - \sqrt{1 + 4\left(\dfrac{1}{2}\right)^{-\frac{\sqrt{2}}{2}}}}{4} \quad , \quad y_6 = 1 + \sqrt{1 + 4\left(\dfrac{1}{2}\right)^{-\frac{\sqrt{2}}{2}}} \quad .$$

(5) : A Composite-Multivariate Dinbakish Equation

$$a_6 (xy)^{2\sqrt{3} - \frac{6}{5}} + a_5 (xy)^{2\sqrt{3}} + a_4 (xy)^{\sqrt{3} - \frac{6}{5}} + a_3 (xy)^{\sqrt{3}} + a_2 (xy)^{-\frac{6}{5}} + a_1 = 0 \ , \ y = a_2 x + a_1 \ ;$$

One of the forms of this Composite-Multivariate Dinbakish Equation can be:

$$\left[a_6 (xy)^{2\sqrt{3}} - 4a_6 (xy)^{\sqrt{3}} - 5a_6 \right]\left[(xy)^{-\frac{6}{5}} - \frac{7}{a_6} \right] = 0 \ ,$$

It expands:

$$a_6 (xy)^{2\sqrt{3} - \frac{6}{5}} - 7(xy)^{2\sqrt{3}} - 4a_6 (xy)^{\sqrt{3} - \frac{6}{5}} + 28(xy)^{\sqrt{3}} - 5a_6 (xy)^{-\frac{6}{5}} + 35 = 0 \ ,$$

And $a_5 = -7; a_4 = -4a_6; a_3 = 28; a_2 = -5a_6; a_1 = 35$.

The Composite-Multivariate Dinbakish Equation follows the real number a_6 to be changed.

When $a_6 = 2$, it means that:

$$a_4 = -8; a_2 = -10 \ .$$

Now, the Composite-Multivariate Dinbakish Equation is:

$$2(xy)^{2\sqrt{3} - \frac{6}{5}} - 7(xy)^{2\sqrt{3}} - 8(xy)^{\sqrt{3} - \frac{6}{5}} + 28(xy)^{\sqrt{3}} - 10(xy)^{-\frac{6}{5}} + 35 = 0 \ , \ y = -10x + 35 \ ;$$

In which it has three parts:

Part 1: $xy = 5^{\frac{\sqrt{3}}{3}}$; Part 2: $xy = (-1)^{\frac{\sqrt{3}}{3}}$; Part 3: $xy = \left(\dfrac{7}{2}\right)^{-\frac{5}{6}}$.

In the part 1:

$$xy = 5^{\frac{\sqrt{3}}{3}} \ , \ y = -10x + 35 \ ;$$

$$y = \frac{5^{\frac{\sqrt{3}}{3}}}{x} \ , \ \left(x - \frac{7}{4}\right)^2 = \frac{49}{16} - \frac{5^{\frac{\sqrt{3}}{3}}}{10} \ ;$$

$$x_1 = \frac{7}{4} + \sqrt{\frac{49}{16} - \frac{5^{\frac{\sqrt{3}}{3}}}{10}} \ , \ y_1 = \frac{35}{2} - 10\sqrt{\frac{49}{16} - \frac{5^{\frac{\sqrt{3}}{3}}}{10}} \ ;$$

$$x_2 = \frac{7}{4} - \sqrt{\frac{49}{16} - \frac{5^{\frac{\sqrt{3}}{3}}}{10}} \ , \ y_2 = \frac{35}{2} + 10\sqrt{\frac{49}{16} - \frac{5^{\frac{\sqrt{3}}{3}}}{10}} \ .$$

In the part 2:

$$xy = (-1)^{\frac{\sqrt{3}}{3}} \ , \ y = -10x + 35 \ ;$$

$$y = \frac{(-1)^{\frac{\sqrt{3}}{3}}}{x} \ , \ \left(x - \frac{7}{4}\right)^2 = \frac{49}{16} - \frac{(-1)^{\frac{\sqrt{3}}{3}}}{10} \ ;$$

$$x_3 = \frac{7}{4} + \sqrt{\frac{49}{16} - \frac{(-1)^{\frac{\sqrt{3}}{3}}}{10}} \ , \ y_3 = \frac{35}{2} - 10\sqrt{\frac{49}{16} - \frac{(-1)^{\frac{\sqrt{3}}{3}}}{10}} \ ;$$

$$x_4 = \frac{7}{4} - \sqrt{\frac{49}{16} - \frac{(-1)^{\frac{\sqrt{3}}{3}}}{10}} \ , \ y_4 = \frac{35}{2} + 10\sqrt{\frac{49}{16} - \frac{(-1)^{\frac{\sqrt{3}}{3}}}{10}} \ .$$

In the part 3:

$$xy = \left(\frac{7}{2}\right)^{-\frac{5}{6}} \ , \ y = -10x + 35 \ ;$$

$$y = \frac{\left(\frac{7}{2}\right)^{-\frac{5}{6}}}{x} \quad , \quad \left(x - \frac{7}{4}\right)^2 = \frac{49}{16} - \frac{\left(\frac{7}{2}\right)^{-\frac{5}{6}}}{10} \; ;$$

$$x_5 = \frac{7}{4} + \sqrt{\frac{49}{16} - \frac{\left(\frac{7}{2}\right)^{-\frac{5}{6}}}{10}} \quad , \quad y_5 = \frac{35}{2} - 10\sqrt{\frac{49}{16} - \frac{\left(\frac{7}{2}\right)^{-\frac{5}{6}}}{10}} \; ;$$

$$x_6 = \frac{7}{4} - \sqrt{\frac{49}{16} - \frac{\left(\frac{7}{2}\right)^{-\frac{5}{6}}}{10}} \quad , \quad y_6 = \frac{35}{2} + 10\sqrt{\frac{49}{16} - \frac{\left(\frac{7}{2}\right)^{-\frac{5}{6}}}{10}} \; .$$

(6) : A Composite-Multivariate Dinbakish Equation

$$a_6 (xy)^{-6\sqrt{10} - \frac{1}{3}} + a_5 (xy)^{-6\sqrt{10}} + a_4 (xy)^{-3\sqrt{10} - \frac{1}{3}} + a_3 (xy)^{-3\sqrt{10}} + a_2 (xy)^{-\frac{1}{3}} + a_1 = 0 \; , \quad y = a_2 x + a_1 \; ;$$

One of the forms of this Composite-Multivariate Dinbakish Equation can be:

$$\left[a_6 (xy)^{-6\sqrt{10}} + \frac{a_6}{2} (xy)^{-3\sqrt{10}} + \frac{2}{3a_6} \right]\left[(xy)^{-\frac{1}{3}} - 3 \right] = 0 \; ,$$

It expands:

$$a_6 (xy)^{-6\sqrt{10} - \frac{1}{3}} - 3a_6 (xy)^{-6\sqrt{10}} + \frac{a_6}{2} (xy)^{-3\sqrt{10} - \frac{1}{3}} - \frac{3a_6}{2} (xy)^{-3\sqrt{10}} + \frac{2}{3a_6} (xy)^{-\frac{1}{3}} - \frac{2}{a_6} = 0 \; ,$$

And $a_5 = -3a_6; a_4 = \frac{a_6}{2}; a_3 = -\frac{3a_6}{2}; a_2 = \frac{2}{3a_6}; a_1 = -\frac{2}{a_6}$.

The Composite-Multivariate Dinbakish Equation follows the real number a_6 to be changed.

When $a_6 = -1$, it means that:

$$a_5 = 3; a_4 = -\frac{1}{2}; a_3 = \frac{3}{2}; a_2 = -\frac{2}{3}; a_1 = 2 \; .$$

Now, the Composite-Multivariate Dinbakish Equation is:

$$-(xy)^{-6\sqrt{10}-\frac{1}{3}} + 3(xy)^{-6\sqrt{10}} - \frac{1}{2}(xy)^{-3\sqrt{10}-\frac{1}{3}} + \frac{3}{2}(xy)^{-3\sqrt{10}} - \frac{2}{3}(xy)^{-\frac{1}{3}} + 2 = 0 \;,\; y = -\frac{2}{3}x + 2 \;.$$

In which it has three parts:

$$\text{Part 1: } xy = \left(-\frac{1}{4} + \sqrt{-\frac{29}{48}}\right)^{-\frac{\sqrt{10}}{30}} \;;\; \text{Part 2: } xy = \left(-\frac{1}{4} - \sqrt{-\frac{29}{48}}\right)^{-\frac{\sqrt{10}}{30}} \;;$$

$$\text{Part 3: } xy = 3^{-3} \;.$$

In the part 1:

$$xy = \left(-\frac{1}{4} + \sqrt{-\frac{29}{48}}\right)^{-\frac{\sqrt{10}}{30}} \;,\; y = -\frac{2}{3}x + 2 \;;$$

$$y = \frac{\left(-\frac{1}{4} + \sqrt{-\frac{29}{48}}\right)^{-\frac{\sqrt{10}}{30}}}{x} \;,\; \left(x - \frac{3}{2}\right)^2 = \frac{9 - 6\left(-\frac{1}{4} + \sqrt{-\frac{29}{48}}\right)^{-\frac{\sqrt{10}}{30}}}{4} \;;$$

$$x_1 = \frac{3 + \sqrt{9 - 6\left(-\frac{1}{4} + \sqrt{-\frac{29}{48}}\right)^{-\frac{\sqrt{10}}{30}}}}{2} \;,\; y_1 = \frac{-3 - \sqrt{9 - 6\left(-\frac{1}{4} + \sqrt{-\frac{29}{48}}\right)^{-\frac{\sqrt{10}}{30}}}}{3} + 2 \;;$$

$$x_2 = \frac{3 - \sqrt{9 - 6\left(-\frac{1}{4} + \sqrt{-\frac{29}{48}}\right)^{-\frac{\sqrt{10}}{30}}}}{2} \;,\; y_2 = \frac{-3 + \sqrt{9 - 6\left(-\frac{1}{4} + \sqrt{-\frac{29}{48}}\right)^{-\frac{\sqrt{10}}{30}}}}{3} + 2 \;.$$

In the part 2:

$$xy = \left(-\frac{1}{4} - \sqrt{-\frac{29}{48}}\right)^{-\frac{\sqrt{10}}{30}} \;,\; y = -\frac{2}{3}x + 2 \;;$$

$$y = \frac{\left(-\frac{1}{4}-\sqrt{-\frac{29}{48}}\right)^{-\frac{\sqrt{10}}{30}}}{x} \quad , \quad \left(x-\frac{3}{2}\right)^2 = \frac{9-6\left(-\frac{1}{4}-\sqrt{-\frac{29}{48}}\right)^{-\frac{\sqrt{10}}{30}}}{4} \quad .$$

$$x_3 = \frac{3+\sqrt{9-6\left(-\frac{1}{4}-\sqrt{-\frac{29}{48}}\right)^{-\frac{\sqrt{10}}{30}}}}{2} \quad , \quad y_3 = \frac{-3-\sqrt{9-6\left(-\frac{1}{4}-\sqrt{-\frac{29}{48}}\right)^{-\frac{\sqrt{10}}{30}}}}{3}+2 \quad ;$$

$$x_4 = \frac{3-\sqrt{9-6\left(-\frac{1}{4}-\sqrt{-\frac{29}{48}}\right)^{-\frac{\sqrt{10}}{30}}}}{2} \quad , \quad y_4 = \frac{-3+\sqrt{9-6\left(-\frac{1}{4}-\sqrt{-\frac{29}{48}}\right)^{-\frac{\sqrt{10}}{30}}}}{3}+2 \quad .$$

In the part 3:

$$xy = 3^{-3} \quad , \quad y = -\frac{2}{3}x+2 \quad ;$$

$$y = \frac{3^{-3}}{x} \quad , \quad \left(x-\frac{3}{2}\right)^2 = \frac{9-6\times3^{-3}}{4} \quad ;$$

$$x_5 = \frac{3+\sqrt{9-6\times3^{-3}}}{2} \quad , \quad y_5 = 1-\frac{\sqrt{9-6\times3^{-3}}}{3} \quad ;$$

$$x_6 = \frac{3-\sqrt{9-6\times3^{-3}}}{2} \quad , \quad y_6 = 1+\frac{\sqrt{9-6\times3^{-3}}}{3} \quad .$$

❷： The other Composite-Multivariate Dinbakish Equations

$$[1]: \quad a_{2n}(xyz)^{(n-1)\frac{i}{k}+\frac{\sqrt{j}}{k}} + a_{2n-1}(xyz)^{(n-1)\frac{i}{k}} + a_{2n-2}(xyz)^{(n-2)\frac{i}{k}+\frac{\sqrt{j}}{k}} + a_{2n-3}(xy)^{(n-2)\frac{i}{k}} + ... + a_3(xyz)^{\frac{i}{k}} + a_2(xyz)^{\frac{\sqrt{j}}{k}} + a_1 = 0 \quad ;$$

$$[2]: \quad a_{2n}(xyz)^{(n-1)\frac{i}{k}+\frac{\sqrt{j}}{m}} + a_{2n-1}(xyz)^{(n-1)\frac{i}{k}} + a_{2n-2}(xyz)^{(n-2)\frac{i}{k}+\frac{\sqrt{j}}{m}} + a_{2n-3}(xy)^{(n-2)\frac{i}{k}} + ... + a_3(xyz)^{\frac{i}{k}} + a_2(xyz)^{\frac{\sqrt{j}}{m}} + a_1 = 0 \quad ;$$

$[3]$: $a_{2n}(xyz)^{(n-1)\frac{i}{k}-\frac{\sqrt{j}}{k}}+a_{2n-1}(xyz)^{(n-1)\frac{i}{k}}+a_{2n-2}(xyz)^{(n-2)\frac{i}{k}-\frac{\sqrt{j}}{k}}+a_{2n-3}(xy)^{(n-2)\frac{i}{k}}+...+a_{3}(xyz)^{\frac{i}{k}}+a_{2}(xyz)^{-\frac{\sqrt{j}}{k}}+a_{1}=0$;

$[4]$: $a_{2n}(xyz)^{(n-1)(-\frac{i}{k})+\frac{\sqrt{j}}{k}}+a_{2n-1}(xyz)^{(n-1)(-\frac{i}{k})}+a_{2n-2}(xyz)^{(n-2)(-\frac{i}{k})+\frac{\sqrt{j}}{k}}+a_{2n-3}(xy)^{(n-2)(-\frac{i}{k})}+...+a_{3}(xyz)^{-\frac{i}{k}}+a_{2}(xyz)^{\frac{\sqrt{j}}{k}}+a_{1}=0$;

$[5]$: $a_{2n}(xyz)^{(n-1)(-\frac{i}{k})-\frac{\sqrt{j}}{k}}+a_{2n-1}(xyz)^{(n-1)(-\frac{i}{k})}+a_{2n-2}(xyz)^{(n-2)(-\frac{i}{k})-\frac{\sqrt{j}}{k}}+a_{2n-3}(xy)^{(n-2)(-\frac{i}{k})}+...+a_{3}(xyz)^{-\frac{i}{k}}+a_{2}(xyz)^{-\frac{\sqrt{j}}{k}}+a_{1}=0$;

$[6]$: $a_{2n}(xyz)^{(n-1)(-\frac{i}{k})+\frac{\sqrt{j}}{m}}+a_{2n-1}(xyz)^{(n-1)(-\frac{i}{k})}+a_{2n-2}(xyz)^{(n-2)(-\frac{i}{k})+\frac{\sqrt{j}}{m}}+a_{2n-3}(xy)^{(n-2)(-\frac{i}{k})}+...+a_{3}(xyz)^{-\frac{i}{k}}+a_{2}(xyz)^{\frac{\sqrt{j}}{m}}+a_{1}=0$;

$[7]$: $a_{2n}(xyz)^{(n-1)(-\frac{i}{k})-\frac{\sqrt{j}}{m}}+a_{2n-1}(xyz)^{(n-1)(-\frac{i}{k})}+a_{2n-2}(xyz)^{(n-2)(-\frac{i}{k})-\frac{\sqrt{j}}{m}}+a_{2n-3}(xy)^{(n-2)(-\frac{i}{k})}+...+a_{3}(xyz)^{-\frac{i}{k}}+a_{2}(xyz)^{-\frac{\sqrt{j}}{m}}+a_{1}=0$;

$[8]$: $a_{2n}(xyz)^{(n-1)\frac{i}{k}-\frac{\sqrt{j}}{m}}+a_{2n-1}(xyz)^{(n-1)\frac{i}{k}}+a_{2n-2}(xyz)^{(n-2)\frac{i}{k}-\frac{\sqrt{j}}{m}}+a_{2n-3}(xy)^{(n-2)\frac{i}{k}}+...+a_{3}(xyz)^{\frac{i}{k}}+a_{2}(xyz)^{-\frac{\sqrt{j}}{m}}+a_{1}=0$;

$[9]$: $a_{2n}(xyz)^{(n-1)\frac{i}{k}+\sqrt{j}}+a_{2n-1}(xyz)^{(n-1)\frac{i}{k}}+a_{2n-2}(xyz)^{(n-2)\frac{i}{k}+\sqrt{j}}+a_{2n-3}(xyz)^{(n-2)\frac{i}{k}}+...+a_{3}(xyz)^{\frac{i}{k}}+a_{2}(xyz)^{\sqrt{j}}+a_{1}=0$;

$[10]$: $a_{2n}(xyz)^{(n-1)\frac{i}{k}-\sqrt{j}}+a_{2n-1}(xyz)^{(n-1)\frac{i}{k}}+a_{2n-2}(xyz)^{(n-2)\frac{i}{k}-\sqrt{j}}+a_{2n-3}(xyz)^{(n-2)\frac{i}{k}}+...+a_{3}(xyz)^{\frac{i}{k}}+a_{2}(xyz)^{-\sqrt{j}}+a_{1}=0$;

$[11]$: $a_{2n}(xyz)^{(n-1)(-\frac{i}{k})-\sqrt{j}}+a_{2n-1}(xyz)^{(n-1)(-\frac{i}{k})}+a_{2n-2}(xyz)^{(n-2)(-\frac{i}{k})-\sqrt{j}}+a_{2n-3}(xyz)^{(n-2)(-\frac{i}{k})}+...+a_{3}(xyz)^{-\frac{i}{k}}+a_{2}(xyz)^{-\sqrt{j}}+a_{1}=0$;

$[12]$: $a_{2n}(xyz)^{(n-1)(-\frac{i}{k})+\sqrt{j}}+a_{2n-1}(xyz)^{(n-1)(-\frac{i}{k})}+a_{2n-2}(xyz)^{(n-2)(-\frac{i}{k})+\sqrt{j}}+a_{2n-3}(xyz)^{(n-2)(-\frac{i}{k})}+...+a_{3}(xyz)^{-\frac{i}{k}}+a_{2}(xyz)^{\sqrt{j}}+a_{1}=0$;

$[13]$: $a_{2n}(xyz)^{(n-1)\sqrt{i}+\frac{\sqrt{j}}{k}}+a_{2n-1}(xyz)^{(n-1)\sqrt{i}}+a_{2n-2}(xyz)^{(n-2)\sqrt{i}+\frac{\sqrt{j}}{k}}+a_{2n-3}(xyz)^{(n-2)\sqrt{i}}+...+a_{3}(xyz)^{\sqrt{i}}+a_{2}(xyz)^{\frac{\sqrt{j}}{k}}+a_{1}=0$;

$[14]$: $a_{2n}(xyz)^{(n-1)\sqrt{i}-\frac{\sqrt{j}}{k}}+a_{2n-1}(xyz)^{(n-1)\sqrt{i}}+a_{2n-2}(xyz)^{(n-2)\sqrt{i}-\frac{\sqrt{j}}{k}}+a_{2n-3}(xyz)^{(n-2)\sqrt{i}}+...+a_{3}(xyz)^{\sqrt{i}}+a_{2}(xyz)^{-\frac{\sqrt{j}}{k}}+a_{1}=0$;

$[15]$: $a_{2n}(xyz)^{(n-1)(-\sqrt{i})+\frac{\sqrt{j}}{k}}+a_{2n-1}(xyz)^{(n-1)(-\sqrt{i})}+a_{2n-2}(xyz)^{(n-2)(-\sqrt{i})+\frac{\sqrt{j}}{k}}+a_{2n-3}(xyz)^{(n-2)(-\sqrt{i})}+...+a_{3}(xyz)^{-\sqrt{i}}+a_{2}(xyz)^{\frac{\sqrt{j}}{k}}+a_{1}=0$;

$[16]$: $a_{2n}(xyz)^{(n-1)(-\sqrt{i})-\frac{\sqrt{j}}{k}}+a_{2n-1}(xyz)^{(n-1)(-\sqrt{i})}+a_{2n-2}(xyz)^{(n-2)(-\sqrt{i})-\frac{\sqrt{j}}{k}}+a_{2n-3}(xyz)^{(n-2)(-\sqrt{i})}+...+a_{3}(xyz)^{-\sqrt{i}}+a_{2}(xyz)^{\frac{\sqrt{j}}{k}}+a_{1}=0$;

$[17]$: $a_{2n}(xyz)^{(n-1)\sqrt{i}+\sqrt{j}}+a_{2n-1}(xyz)^{(n-1)\sqrt{i}}+a_{2n-2}(xyz)^{(n-2)\sqrt{i}+\sqrt{j}}+a_{2n-3}(xyz)^{(n-2)\sqrt{i}}+...+a_{3}(xyz)^{\sqrt{i}}+a_{2}(xyz)^{\sqrt{j}}+a_{1}=0$;

$[18]$: $a_{2n}(xyz)^{(n-1)\sqrt{i}-\sqrt{j}}+a_{2n-1}(xyz)^{(n-1)\sqrt{i}}+a_{2n-2}(xyz)^{(n-2)\sqrt{i}-\sqrt{j}}+a_{2n-3}(xyz)^{(n-2)\sqrt{i}}+...+a_{3}(xyz)^{\sqrt{i}}+a_{2}(xyz)^{-\sqrt{j}}+a_{1}=0$;

$[19]:$ $a_{2n}(xyz)^{(n-1)(-\sqrt{i})+\sqrt{j}}+a_{2n-1}(xyz)^{(n-1)(-\sqrt{i})}+a_{2n-2}(xyz)^{(n-2)(-\sqrt{i})+\sqrt{j}}+a_{2n-3}(xyz)^{(n-2)(-\sqrt{i})}+...+a_3(xyz)^{-\sqrt{i}}+a_2(xyz)^{\sqrt{j}}+a_1=0$;

$[20]:$ $a_{2n}(xyz)^{(n-1)(-\sqrt{i})-\sqrt{j}}+a_{2n-1}(xyz)^{(n-1)(-\sqrt{i})}+a_{2n-2}(xyz)^{(n-2)(-\sqrt{i})-\sqrt{j}}+a_{2n-3}(xyz)^{(n-2)(-\sqrt{i})}+...+a_3(xyz)^{-\sqrt{i}}+a_2(xyz)^{-\sqrt{j}}+a_1=0$.

$(i=2,3,4,...,\infty; m=2,3,4,...,\infty; j=2,3,4,...,\infty; n=2,3,4,...,\infty; k=2,3,4,...,\infty)$, $y=\sqrt{x}+a_1; z=\sqrt{x}-a_1$.

… …

About the Complex Composite-Multivariate Dinbakish Equations, I have no time to show them in the Volume 2. Also in fact, they have other things in which they are always great.

Part 2

1: The Dinbakish Theorems

(1) : An Equation

$$ax^5+bx^4+cx^3+dx^2+ex+f=0$$

$(a \in R, b \in R, c \in R, d \in R, e \in R, f \in R, a \neq 0, b \neq 0, c \neq 0, d \neq 0, e \neq 0, f \neq 0)$

i One of the forms of this equation can be:

$$\left(ax-5\right)\left(x-\frac{2}{a}\right)\left(x+\frac{2}{a}\right)(x-3a)(x+4a)=0 ,$$

It expands:

$$ax^5+\left(a^2-5\right)x^4-\left(5a+12a^3+\frac{4}{a}\right)x^3+\left(60a^2+\frac{20}{a^2}-4\right)x^2+\left(\frac{20}{a}+48a\right)x-240=0 ,$$

And $b = a^2 - 5; c = -\left(5a + 12a^3 + \dfrac{4}{a}\right); d = 60a^2 + \dfrac{20}{a^2} - 4; e = \dfrac{20}{a} + 48a; f = -240$.

The equation follows the real number a to be changed.

1: when $a = -4$, it means that:

$$b = 11; c = 789; d = \dfrac{3829}{4}; e = 197 .$$

Now, the equation is:

$$-4x^5 + 11x^4 + 789x^3 + \dfrac{3829}{4}x^2 - 197x - 240 = 0 ,$$

Its roots are: 1) $-4x - 5 = 0$, $x_1 = -\dfrac{5}{4}$; 2) $x + \dfrac{1}{2} = 0$, $x_2 = -\dfrac{1}{2}$; 3) $x - \dfrac{1}{2} = 0$, $x_3 = \dfrac{1}{2}$;

$\qquad$ 4) $x + 12 = 0$, $x_4 = -12$; 5) $x - 16 = 0$, $x_5 = 16$.

2: when $a = \dfrac{\sqrt{2}}{2}$, it means that:

$$b = -\dfrac{9}{2}; c = -\dfrac{19\sqrt{2}}{2}; d = 66; e = 44\sqrt{2} .$$

Now, the equation is:

$$\dfrac{\sqrt{2}}{2}x^5 - \dfrac{9}{2}x^4 - \dfrac{19\sqrt{2}}{2}x^3 + 66x^2 + 44\sqrt{2}x - 240 = 0 ,$$

Its roots are: 1) $\dfrac{\sqrt{2}}{2}x - 5 = 0$, $x_1 = 5\sqrt{2}$; 2) $x - 2\sqrt{2} = 0$, $x_2 = 2\sqrt{2}$; 3) $x + 2\sqrt{2} = 0$, $x_3 = -2\sqrt{2}$;

$\qquad$ 4) $x - \dfrac{3\sqrt{2}}{2} = 0$, $x_4 = \dfrac{3\sqrt{2}}{2}$; 5) $x + 2\sqrt{2} = 0$, $x_5 = -2\sqrt{2}$.

$$\dots\ \dots\infty\ \dots\ \dots$$

So that when $a \in R, a \neq 0, b = a^2 - 5, c = -\left(5a + 12a^3 + \dfrac{4}{a}\right)$, $d = 60a^2 + \dfrac{20}{a^2} - 4, e = \dfrac{20}{a} + 48a$

$f = -240$, the equation $ax^5 + bx^4 + cx^3 + dx^2 + ex + f = 0$, which has the roots: $x_1 = \dfrac{5}{a}$,

$x_2 = \dfrac{2}{a}, x_3 = -\dfrac{2}{a}, x_4 = 3a, x_5 = -4a$. I call it one of the Dinbakish Theorems.

①: According to the equation $ax^5 + bx^4 + cx^3 + dx^2 + ex + f = 0$, can you fill in the blanks with the correct numbers for it?

[1]: $(\ ) + (-528) + (\ ) + (\ ) + (\ ) + (-6) = 0$;

[2]: $(\ ) + (\ ) + (\ ) + (\ ) + (178) + (-6) = 0$;

[3]: $\left(\dfrac{1}{8}\right) + (\ ) + (\ ) + (\ ) + (\ ) + (-6) = 0$;

[4]: $(\ ) + (\ ) + (\ ) + \left(-\dfrac{411}{64}\right) + (\ ) + (-6) = 0$;

[5]: $(\ ) + (\ ) + \left(\dfrac{2457}{8}\right) + (\ ) + (\ ) + (-6) = 0$;

$a = (\ ); b = (\ ); c = (\ ); d = (\ ); e = (\ );\ x_1 = (\ ); x_2 = (\ ); x_3 = (\ ); x_4 = (\ ); x_5 = (\ )$.

ii: One of the forms of this equation can be:

$$\left(ax^2 + 2ax + a\right)\left(x - \dfrac{3}{a}\right)\left(x + \dfrac{3}{a}\right)(x - 5) = 0 ,$$

It expands:

$$ax^5 - 3ax^4 - \left(9a + \dfrac{9}{a}\right)x^3 - \left(5a - \dfrac{27}{a}\right)x^2 + \dfrac{81}{a}x + \dfrac{45}{a} = 0 ,$$

And $b = -3a; c = -\left(9a + \dfrac{9}{a}\right); d = -\left(5a - \dfrac{27}{a}\right); e = \dfrac{81}{a}; f = \dfrac{45}{a}$.

The equation follows the real number a to be changed.

74

❶: when $a=3$, it means that:

$b=-9; c=-30; d=-6; e=27; f=15$.

Now, the equation is:

$3x^5 - 9x^4 - 30x^3 - 6x^2 + 27x + 15 = 0$,

Its roots are: 1) $x^2 + 2x + 1 = 0$, $x_1 = x_2 = -1$; 2) $x - 1 = 0$, $x_3 = 1$; 3) $x + 1 = 0$, $x_4 = -1$;

4) $x - 5 = 0$, $x_5 = 5$.

❷: when $a = -\dfrac{1}{3}$, it means that:

$b = 1; c = 30; d = -\dfrac{238}{3}; e = -243; f = -135$.

Now, the equation is:

$-\dfrac{1}{3}x^5 + x^4 + 30x^3 - \dfrac{238}{3}x^2 - 243x - 135 = 0$,

Its roots are: 1) $-\dfrac{1}{3}x^2 - \dfrac{2}{3}x - \dfrac{1}{3} = 0$, $x_1 = x_2 = -1$; 2) $x + 9 = 0$, $x_3 = -9$;

3) $x - 9 = 0$, $x_4 = 9$; 4) $x - 5 = 0$, $x_5 = 5$.

$\dots \dots \infty \dots \dots$

So that when $a \in R; a \neq 0; b = -3a; c = -\left(9a + \dfrac{9}{a}\right); d = -\left(5a - \dfrac{27}{a}\right); e = \dfrac{81}{a}; f = \dfrac{45}{a}$, the equation

$ax^5 + bx^4 + cx^3 + dx^2 + ex + f = 0$, which has the roots: $x_1 = x_2 = -1; x_3 = \dfrac{3}{a}; x_4 = -\dfrac{3}{a}; x_5 = 5$.
I call it one of the Dinbakish Theorems.

②: According to the equation $ax^5 + bx^4 + cx^3 + dx^2 + ex + f = 0$, can you fill in the blanks with the correct numbers for it?

$[1]$: $(\ )+(\ )+(\ )+(\ )+(16)+(-24)=0$;

$[2]$: $(-32)+(\ )+(\ )+(\ )+(\ )+(-24)=0$;

$[3]$: $(\ )+(\ )+(-162)+(\ )+(\ )+(-24)=0$;

$[4]$: $(\ )+(\ )+(\ )+(36)+(\ )+(-24)=0$;

$[5]$: $(\ )+(-12)+(\ )+(\ )+(\ )+(-24)=0$;

$a=(\ );b=(\ );c=(\ );d=(\ );e=(\ );x_1=(\ );x_2=(\ );x_3=(\ );x_4=(\ );x_5=(\ )$.

iii: One of the forms of this equation can be:

$$\left(ax^2+4ax+3\right)\left(x^2+5x+a\right)(x-a)=0 \ ,$$

It expands:

$$ax^5-\left(a^2-9a\right)x^4-\left(8a^2-20a-3\right)x^3-\left(a^3+16a^2+3a-15\right)x^2-\left(4a^3+12a\right)x-3a^2=0 \ ,$$

And $b=-\left(a^2-9a\right);c=-\left(8a^2-20a-3\right);d=-\left(a^3+16a^2+3a-15\right);e=-\left(4a^3+12a\right);f=-3a^2$.

The equation follows the real number a to changed.

❶: when $a=1$, it means that:

$b=8;c=15;d=-5;e=-16;f=-3$.

Now, the equation is:

$x^5+8x^4+15x^3-5x^2-16x-3=0$,

Its roots are: 1) $x^2+4x+3=0$, $x_1=-1;x_2=-3$;

2) $x^2+5x+1=0$, $x_3=\dfrac{-5+\sqrt{21}}{2};x_4=\dfrac{-5-\sqrt{21}}{2}$; 3) $x-1=0$, $x_5=1$.

❷: when $a=\sqrt{3}$, it means that:

$b=-\left(3-9\sqrt{3}\right);c=-\left(21-20\sqrt{3}\right);d=-\left(6\sqrt{3}+33\right);e=-24\sqrt{3};f=-9$.

Now, the equation is:

$$\sqrt{3}x^5-\left(3-9\sqrt{3}\right)x^4-\left(21-20\sqrt{3}\right)x^3-\left(6\sqrt{3}+33\right)x^2-24\sqrt{3}x-9=0 \; ;$$

Its roots are: 1) $\sqrt{3}x^2+4\sqrt{3}x+3=0$, $x_1=-2+\sqrt{4-\sqrt{3}};x_2=-2-\sqrt{4-\sqrt{3}};$

2) $x^2+5x+\sqrt{3}=0$, $x_3=-\dfrac{5}{2}+\sqrt{\dfrac{25}{4}-\sqrt{3}};x_4=-\dfrac{5}{2}-\sqrt{\dfrac{25}{4}-\sqrt{3}}$; 3) $x-\sqrt{3}=0$, $x_5=\sqrt{3}$.

$$\dots \dots \infty \dots$$

So that when $a\in R;a\neq0;b=-\left(a^2-9a\right);c=-\left(8a^2-20a-3\right);d=-\left(a^3+16a^2+3a-15\right);e=-\left(4a^3+12a\right);f=-3a^2$,

the equation $ax^5+bx^4+cx^3+dx^2+ex+f=0$, which has the roots: $x_1=-2+\sqrt{4-\dfrac{3}{a}};x_2=-2-\sqrt{4-\dfrac{3}{a}}$;

$x_3=-\dfrac{5}{2}+\sqrt{\dfrac{25}{4}-a};x_4=-\dfrac{5}{2}-\sqrt{\dfrac{25}{4}-a}$; $x_5=a$. I call it one of the Dinbakish Theorems.

③: According to the equation $ax^5+bx^4+cx^3+dx^2+ex+f=0$, can you fill in the blanks with the correct numbers for it?

[1]: $(-243)+(\;\;)+(\;\;)+(\;\;)+(\;\;)+(-54)=0$;

[2]: $(\;\;)+(\;\;)+(\;\;)+(\;\;)+(81)+(-54)=0$;

[3]: $(\;\;)+(\;\;)+(-96)+(\;\;)+(\;\;)+(-54)=0$;

[4]: $(\;\;)+(-18)+(\;\;)+(\;\;)+(\;\;)+(-54)=0$;

[5]: $(\;\;)+(\;\;)+(\;\;)+(48)+(\;\;)+(-54)=0$;

$a=(\;\;);b=(\;\;);c=(\;\;);d=(\;\;);e=(\;\;);x_1=(\;\;);x_2=(\;\;);x_3=(\;\;);x_4=(\;\;);x_5=(\;\;)$.

$$(2) : \text{An Equation}$$

$$ax^6 + bx^5 + cx^4 + dx^3 + ex^2 + fx + g = 0$$

$$(a \in R; b \in R; c \in R; d \in R; e \in R; f \in R; g \in R; a \ne 0; b \ne 0; c \ne 0; d \ne 0; e \ne 0; f \ne 0; g \ne 0)$$

i. One of the forms of this equation can be:

$$a\left(x - \sqrt{3}\right)\left(x + \sqrt{3}\right)\left(x - \sqrt{6}\right)\left(x + \sqrt{6}\right)\left(x - 5\right)\left(x - \frac{3}{2a}\right) = 0 ,$$

It expands:

$$ax^6 - \left(\frac{3}{2} + 5a\right)x^5 + \left(\frac{15}{2} - 9a\right)x^4 + \left(45a + \frac{27}{2}\right)x^3 - \left(\frac{135}{2} - 18a\right)x^2 - \left(90a + 27\right)x + 135 = 0 ,$$

And $b = -\left(\dfrac{3}{2} + 5a\right); c = \dfrac{15}{2} - 9a; d = 45a + \dfrac{27}{2}; e = -\left(\dfrac{135}{2} - 18a\right); f = -\left(90a + 27\right); g = 135$.

The equation follows the real number a to be changed.

1 : when $a = \dfrac{3}{2}$, it means that:

$$b = -9; c = -6; d = 81; e = -\frac{81}{2}; f = -162 .$$

Now, the equation is:

$$\frac{3}{2}x^6 - 9x^5 - 6x^4 + 81x^3 - \frac{81}{2}x^2 - 162x + 135 = 0 ,$$

Its roots are: 1) $x - \sqrt{3} = 0, x_1 = \sqrt{3}$; 2) $x + \sqrt{3} = 0, x_2 = -\sqrt{3}$; 3) $x - \sqrt{6} = 0, x_3 = \sqrt{6}$;

4) $x + \sqrt{6} = 0, x_4 = -\sqrt{6}$; 5) $x - 5 = 0, x_5 = 5$; 6) $x - 1 = 0, x_6 = 1$.

2 : when $a = 2\sqrt{2}$, it means that:

$$b = -\left(\frac{3}{2} + 10\sqrt{2}\right); c = \frac{15}{2} - 18\sqrt{2}; d = 90\sqrt{2} + \frac{27}{2}; e = -\left(\frac{135}{2} - 36\sqrt{2}\right); f = -\left(180\sqrt{2} + 27\right) .$$

Now, the equation is:

$$2\sqrt{2}x^6 -\left(\frac{3}{2}+10\sqrt{2}\right)x^5 +\left(\frac{15}{2}-18\sqrt{2}\right)x^4 +\left(90\sqrt{2}+\frac{27}{2}\right)x^3 -\left(\frac{135}{2}-36\sqrt{2}\right)x^2 -\left(180\sqrt{2}+27\right)x+135=0 \ ,$$

Its roots are: 1) $x-\sqrt{3}=0, x_1=\sqrt{3}$; 2) $x+\sqrt{3}=0, x_2=-\sqrt{3}$; 3) $x-\sqrt{6}=0, x_3=\sqrt{6}$;

4) $x+\sqrt{6}=0, x_4=-\sqrt{6}$; 5) $x-5=0, x_5=5$; 6) $x-\frac{3\sqrt{2}}{8}=0, x_6=\frac{3\sqrt{2}}{8}$.

$$\ldots \ldots \infty \ldots \ldots$$

So that when $a\in R; a\neq 0; b=-\left(\frac{3}{2}+5a\right); c=\frac{15}{2}-9a; d=45a+\frac{27}{2}; e=-\left(\frac{135}{2}-18a\right); f=-(90a+27); g=135$,

the equation $ax^6+bx^5+cx^4+dx^3+ex^2+fx+g=0$, which has the roots: $x_1=\sqrt{3}; x_2=-\sqrt{3};$

$x_3=\sqrt{6}; x_4=-\sqrt{6}; x_5=5; x_6=\frac{3}{2a}$. I call it one of the Dinbakish Theorems.

①: According to the equation $ax^6+bx^5+cx^4+dx^3+ex^2+fx+g=0$, can you fill in the blanks with the correct numbers for it?

[1]: $\left(\frac{1}{32}\right)+(\ )+(-1)+(\ )+(\ )+(\ )+(36)=0$;

[2]: $(\ )+(-9)+(\ )+(90)+(\ )+(\ )+(36)=0$;

[3]: $(\ )+(\ )+(-16)+(\ )+(\ )+(\ )+(36)=0$;

[4]: $(1458)+(\ )+(\ )+(\ )+(\ )+(-243)+(36)=0$;

[5]: $(\ )+(\ )+(\ )+(\ )+(-198)+(\ )+(36)=0$;

[6]: $(\ )+(-9216)+(\ )+(5760)+(\ )+(\ )+(36)=0$;

$a=(\ ); b=(\ ); c=(\ ); d=(\ ); e=(\ ); f=(\ ); x_1=(\ ); x_2=(\ )$;

$x_3=(\ ); x_4=(\ ); x_5=(\ ); x_6=(\ )$.

ii: One of the forms of this equation can be:

$$\left(ax^2 - ax + 1\right)\left(x - \sqrt{a}\right)\left(x + \sqrt{a}\right)\left(x - \frac{\sqrt{a}}{a}\right)\left(x + \frac{\sqrt{a}}{a}\right) = 0 ,$$

It expands: $ax^6 - ax^5 - a^2x^4 + \left(a^2 + 1\right)x^3 - \frac{1}{a}x^2 - ax + 1 = 0$,

And $b = -a; c = -a^2; d = a^2 + 1; e = -\frac{1}{a}; f = -a; g = 1$,

The equation follows the real number a to be changed.

❶: when $a = -\frac{1}{4}$, it means that:

$$b = f = \frac{1}{4}; c = -\frac{1}{16}; d = \frac{17}{16}; e = 4 .$$

Now, the equation is:

$$-\frac{1}{4}x^6 + \frac{1}{4}x^5 - \frac{1}{16}x^4 + \frac{17}{16}x^3 + 4x^2 + \frac{1}{4}x + 1 = 0 ,$$

Its roots are: 1) $-\frac{1}{4}x^2 + \frac{1}{4}x + 1 = 0; x_1 = \frac{1 + \sqrt{17}}{2}, x_2 = \frac{1 - \sqrt{17}}{2}$; 2) $x - \sqrt{-\frac{1}{4}} = 0, x_3 = \frac{i}{2}$;

3) $x + \sqrt{-\frac{1}{4}} = 0, x_4 = -\frac{i}{2}$; 4) $x - 2i = 0, x_5 = 2i$; 5) $x + 2i = 0, x_6 = -2i$.

❷: when $a = \frac{\sqrt{3}}{3}$, it means that:

$$b = f = -\frac{\sqrt{3}}{3}; c = -\frac{1}{3}; d = \frac{4}{3}; e = -\sqrt{3};$$

Now, the equation is:

$$\frac{\sqrt{3}}{3}x^6 - \frac{\sqrt{3}}{3}x^5 - \frac{1}{3}x^4 + \frac{4}{3}x^3 - \sqrt{3}x^2 - \frac{\sqrt{3}}{3}x + 1 = 0,$$

Its roots are: 1) $\frac{\sqrt{3}}{3}x^2 - \frac{\sqrt{3}}{3}x + 1 = 0, x_1 = \frac{1}{2} + \sqrt{\frac{1}{4} - \sqrt{3}}, x_2 = \frac{1}{2} - \sqrt{\frac{1}{4} - \sqrt{3}}$; 2) $x - \sqrt{\frac{\sqrt{3}}{3}} = 0, x_3 = \sqrt{\frac{\sqrt{3}}{3}}$;

3) $x+\sqrt{\dfrac{\sqrt{3}}{3}}=0, x_4 = -\sqrt{\dfrac{\sqrt{3}}{3}}$; 4) $x - 3^{\frac{1}{4}} = 0, x_5 = 3^{\frac{1}{4}}$; 5) $x + 3^{\frac{1}{4}} = 0, x_6 = -3^{\frac{1}{4}}$.

$$\ldots \ \ldots \infty \ \ldots \ \ldots$$

So when $a \in R; a \neq 0; b = -a; c = -a^2; d = a^2 + 1; e = -\dfrac{1}{a}; f = -a; g = 1$, the equation

$ax^6 + bx^5 + cx^4 + dx^3 + ex^2 + fx + g = 0$, which has the roots: $x_1 = \dfrac{1}{2} + \sqrt{\dfrac{1}{4} - \dfrac{1}{a}}; x_2 = \dfrac{1}{2} - \sqrt{\dfrac{1}{4} - \dfrac{1}{a}}$;

$x_3 = \sqrt{a}; x_4 = -\sqrt{a}; x_5 = \dfrac{\sqrt{a}}{a}; x_6 = -\dfrac{\sqrt{a}}{a}$. I call it one of the Dinbakish Theorems.

②: According to the equation $ax^6 + bx^5 + cx^4 + dx^3 + ex^2 + fx + g = 0$, can you fill in the blanks with the correct numbers for it?

[1]: $(1)+(\ )+(-6)+(\ )+(\ )+(\ )+(-4)=0$;

[2]: $(\ )+(-8)+(\ )+(\ )+(\ )+(-32)+(-4)=0$;

[3]: $(\ )+(\ )+(\ )+(\ )+(1)+(\ )+(-4)=0$;

[4]: $(-64)+(\ )+(\ )+\left(\dfrac{320}{3}\right)+(\ )+(\ )+(-4)=0$;

[5]: $(\ )+\left(\dfrac{256}{3}\right)+(\ )+(\ )+(\ )+(\ )+(-4)=0$;

[6]: $(\ )+(\ )+(\ )+(\ )+(36)+(\ )+(-4)=0$;

$a=(\ ); b=(\ ); c=(\ ); d=(\ ); e=(\ ); f=(\ )$;

$x_1=(\ ); x_2=(\ ); x_3=(\ ); x_4=(\ ); x_5=(\ ); x_6=(\ )$.

iii One of the forms of this equation can be:

$$\left(x^2 + 2ax - 2\right)\left(x^2 - 4ax + 3\right)\left(ax - 2\right)\left(x + \frac{1}{4a}\right) = 0 \ ,$$

It expands:

$$ax^6 - \left(2a^2 + \frac{7}{4}\right)x^5 - \left(8a^3 - \frac{9a}{2} + \frac{1}{2a}\right)x^4 + \left(28a^2 - \frac{3}{4}\right)x^3 - \left(\frac{53a^2 + 1}{2a}\right)x^2 + \frac{7}{2}x + \frac{3}{a} = 0 \ ,$$

And $b = -\left(2a^2 + \frac{7}{4}\right); c = -\left(8a^3 - \frac{9a}{2} + \frac{1}{2a}\right); d = 28a^2 - \frac{3}{4}; e = -\left(\frac{53a^2 + 1}{2a}\right); f = \frac{7}{2}; g = \frac{3}{a}$.

The equation follows the real number a to be changed.

❶ : when $a = -1$, it means that:

$$b = -\frac{15}{4}; c = 4; d = \frac{109}{4}; e = 27; g = -3 \ .$$

Now, the equation is:

$$-x^6 - \frac{15}{4}x^5 + 4x^4 + \frac{109}{4}x^3 + 27x^2 + \frac{7}{2}x - 3 = 0 \ ,$$

Its roots are: 1) $x^2 - 2x - 2 = 0, x_1 = 1 + \sqrt{3}; x_2 = 1 - \sqrt{3}$; 2) $x^2 + 4x + 3 = 0, x_3 = -1; x_4 = -3$;

$$3) \ -x - 2 = 0, x_5 = -2 \ ; 4) \ x - \frac{1}{4} = 0, x_6 = \frac{1}{4} \ .$$

❷ : when $a = \frac{\sqrt{3}}{2}$, it means that:

$$b = -\frac{13}{4}; c = -\frac{7\sqrt{3}}{12}; d = \frac{81}{4}; e = -\frac{163\sqrt{3}}{12}; g = 2\sqrt{3} \ .$$

Now, the equation is:

$$\frac{\sqrt{3}}{2}x^6 - \frac{13}{4}x^5 - \frac{7\sqrt{3}}{12}x^4 + \frac{81}{4}x^3 - \frac{163\sqrt{3}}{12}x^2 + \frac{7}{2}x + 2\sqrt{3} = 0 \ ,$$

Its roots are: 1) $x^2 + \sqrt{3}x - 2 = 0, x_1 = \frac{-\sqrt{3} + \sqrt{11}}{2}; x_2 = \frac{-\sqrt{3} - \sqrt{11}}{2}$;

2) $x^2 - 2\sqrt{3}x + 3 = 0, x_3 = x_4 = \sqrt{3}$; 3) $\dfrac{\sqrt{3}}{2}x - 2 = 0, x_5 = \dfrac{4\sqrt{3}}{3}$; 4) $x + \dfrac{\sqrt{3}}{6} = 0, x_6 = -\dfrac{\sqrt{3}}{6}$.

$$\ldots \ldots \infty \ldots \ldots$$

So that when $a \in R; a \neq 0; b = -\left(2a^2 + \dfrac{7}{4}\right); c = -\left(8a^3 - \dfrac{9a}{2} + \dfrac{1}{2a}\right); d = 28a^2 - \dfrac{3}{4}; e = -\left(\dfrac{53a^2 + 1}{2a}\right); f = \dfrac{7}{2}; g = \dfrac{3}{a}$,

the equation $ax^6 + bx^5 + cx^4 + dx^3 + ex^2 + fx + g = 0$, which has the roots: $x_1 = \dfrac{2}{a}; x_2 = -\dfrac{1}{4a}$;

$x_3 = -a + \sqrt{a^2 + 1}; x_4 = -a - \sqrt{a^2 + 1}$; $x_5 = 2a + \sqrt{4a^2 - 3}; x_6 = 2a - \sqrt{4a^2 - 3}$.

I call it one of the Dinbakish Theorems.

③: According to the equation $ax^6 + bx^5 + cx^4 + dx^3 + ex^2 + fx + g = 0$, can you fill in the blanks with the correct numbers for it?

[1]: $(125) + (\ ) + (\ ) + (\ ) + (\ ) + (\ ) + (60) = 0$;

[2]: $(\ ) + (\ ) + (-125) + (\ ) + (-35) + (\ ) + (60) = 0$;

[3]: $(\ ) + (-4) + (\ ) + (\ ) + (\ ) + (-80) + (60) = 0$;

[4]: $(\ ) + (\ ) + (\ ) + (972) + (-63) + (\ ) + (60) = 0$;

[5]: $(\ ) + (-128) + (-80) + (\ ) + (\ ) + (\ ) + (60) = 0$;

[6]: $(\ ) + (\ ) + (\ ) + (-288) + (\ ) + (160) + (60) = 0$;

$a = (\ ); b = (\ ); c = (\ ); d = (\ ); e = (\ ); f = (\ )$;

$x_1 = (\ ); x_2 = (\ ); x_3 = (\ ); x_4 = (\ ); x_5 = (\ ); x_6 = (\ )$.

iv: One of the forms of this equation can be:

$$\left(ax^2 - 2x - \dfrac{3}{a}\right)\left(x^2 + ax + a\right)\left(x^2 + x - 2a\right) = 0 ,$$

83

It expands:

$$ax^6 + \left(a^2 + a - 2\right)x^5 - \left(2a + 2 + \frac{3}{a}\right)x^4 + \left(a^2 - 2a^3 - \frac{3}{a} - 3\right)x^3 + \left(2a^3 - 4a^2 + 2a\right)x^2 + \left(4a^2 + 6a - 3\right)x + 6a = 0 ,$$

And $b = a^2 + a - 2; c = -\left(2a + 2 + \frac{3}{a}\right); d = a^2 - 2a^3 - \frac{3}{a} - 3; e = 2a^3 - 4a^2 + 2a; f = 4a^2 + 6a - 3; g = 6a$.

The equation follows the real number a to be changed.

❶ : when $a = -3$, it means that:

$d = 4; c = 5; d = 61; e = 96; f = 15; g = -18$.

Now, the equation is:

$$-3x^6 + 4x^5 + 5x^4 + 61x^3 + 96x^2 + 15x - 18 = 0 ,$$

Its roots are: 1) $-3x^2 - 2x + 1 = 0, x_1 = -1; x_2 = -\frac{1}{3}$; 2) $x^2 - 3x - 3 = 0, x_3 = \frac{3 + \sqrt{21}}{2}; x_4 = \frac{3 - \sqrt{21}}{2}$;

3) $x^2 + x + 6 = 0, x_5 = \frac{-1 + \sqrt{23}i}{2}; x_6 = \frac{-1 - \sqrt{23}i}{2}$.

❷ : when $a = \frac{1}{3}$, it means that:

$b = -\frac{14}{9}; c = -\frac{35}{3}; d = -\frac{323}{27}; e = \frac{8}{27}; f = -\frac{5}{9}; g = 2$.

Now, the equation is:

$$\frac{1}{3}x^6 - \frac{14}{9}x^5 - \frac{35}{3}x^4 - \frac{323}{27}x^3 + \frac{8}{27}x^2 - \frac{5}{9}x + 2 = 0 ,$$

Its roots are: 1) $\frac{1}{3}x^2 - 2x - 9 = 0, x_1 = -3; x_2 = 9$;

2) $x^2 + \frac{1}{3}x + \frac{1}{3} = 0, x_3 = \frac{-1 + \sqrt{11}i}{6}; x_4 = \frac{-1 - \sqrt{11}i}{6}$; $x^2 + x - \frac{2}{3} = 0, x_5 = -\frac{1}{2} + \sqrt{\frac{11}{12}}; x_6 = -\frac{1}{2} - \sqrt{\frac{11}{12}}$.

… … ∞ … …

So that when $a \in R; a \neq 0; b = a^2 + a - 2; c = -\left(2a + 2 + \dfrac{3}{a}\right); d = a^2 - 2a^3 - \dfrac{3}{a} - 3;$

$$e = 2a^3 - 4a^2 + 2a; f = 4a^2 + 6a - 3; g = 6a ;$$

the equation $ax^6 + bx^5 + cx^4 + dx^3 + ex^2 + fx + g = 0$, which has the roots: $x_1 = \dfrac{3}{a}; x_2 = -\dfrac{1}{a};$

$$x_3 = -\dfrac{a}{2} + \sqrt{\dfrac{a^2}{4} - a}; x_4 = -\dfrac{a}{2} - \sqrt{\dfrac{a^2}{4} - a} \; ; \; x_5 = -\dfrac{1}{2} + \sqrt{\dfrac{1}{4} + 2a}; x_6 = -\dfrac{1}{2} - \sqrt{\dfrac{1}{4} + 2a} \; .$$

I call it one of the Dinbakish Theorems.

④: According to the equation $ax^6 + bx^5 + cx^4 + dx^3 + ex^2 + fx + g = 0$, can you fill in the blanks with the correct numbers for it?

$[1]$: $(192) + (\quad) + (-912) + (\quad) + (\quad) + (\quad) + (192) = 0$;

$[2]$: $(\quad) + (320) + (\quad) + (\quad) + (528) + (\quad) + (192) = 0$;

$[3]$: $(\quad) + (\quad) + (\quad) + (\quad) + (\quad) + (-2560) + (192) = 0$;

$[4]$: $(\quad) + (\quad) + (\quad) + (-12800) + (\quad) + (\quad) + (192) = 0$;

$[5]$: $(\quad) + (-2430) + (\quad) + (5400) + (\quad) + (\quad) + (192) = 0$;

$[6]$: $\left(\dfrac{1}{243}\right) + (\quad) + (\quad) + (\quad) + (\quad) + \left(-\dfrac{640}{3}\right) + (192) = 0$;

$a = (\quad); b = (\quad); c = (\quad); d = (\quad); e = (\quad); f = (\quad) ;$

$x_1 = (\quad); x_2 = (\quad); x_3 = (\quad); x_4 = (\quad); x_5 = (\quad); x_6 = (\quad) .$

(3) : An Equation

$$ax^7 + bx^6 + cx^5 + dx^4 + ex^3 + fx^2 + gx + h = 0$$

$$(\ a \in R; b \in R; c \in R; d \in R; e \in R; f \in R; g \in R; h \in R;\ )$$

$$(\ a \neq 0; b \neq 0; c \neq 0; d \neq 0; e \neq 0; f \neq 0; g \neq 0; h \neq 0\ )$$

One of the forms of this equation can be:

$$a\left(x^2 - 2a\right)\left(x^2 - \frac{4}{a^2}\right)\left(x^2 - \frac{a}{4}\right)(x - 4) = 0\ ,$$

It expands:

$$ax^7 - 4ax^6 - \left(\frac{9a^2}{4} + \frac{4}{a}\right)x^5 + \left(9a^2 + \frac{16}{a}\right)x^4 + \left(\frac{a^3}{2} + 9\right)x^3 - \left(2a^3 + 36\right)x^2 - 2ax + 8a = 0\ ,$$

And $b = -4a; c = -\left(\dfrac{9a^2}{4} + \dfrac{4}{a}\right); d = 9a^2 + \dfrac{16}{a}; e = \dfrac{a^3}{2} + 9; f = -\left(2a^3 + 36\right); g = -2a; h = 8a$.

The equation follows the real number a to be changed.

❶: when $a = -2$; it means that:

$b = 8; c = -7; d = 28; e = 5; f = -20; g = 4; h = -16$.

Now, the equation is:

$$-2x^7 + 8x^6 - 7x^5 + 28x^4 + 5x^3 - 20x^2 + 4x - 16 = 0\ .$$

Its roots are: 1) $x^2 + 4 = 0, x_1 = 2i; x_2 = -2i$; 2) $x^2 - 1 = 0, x_3 = 1; x_4 = -1$;

3) $x^2 + \dfrac{1}{2} = 0, x_5 = \dfrac{\sqrt{2}i}{2}; x_6 = -\dfrac{\sqrt{2}i}{2}$; 4) $x - 4 = 0, x_7 = 4$.

❷: when $a = \dfrac{1}{5}$, it means that:

$b = -\dfrac{4}{5}; c = -\dfrac{2009}{100}; d = \dfrac{2009}{25}; e = \dfrac{2251}{250}; f = -\dfrac{4502}{125}; g = -\dfrac{2}{5}; h = \dfrac{8}{5}$.

Now, the equation is:

$$\frac{1}{5}x^7 - \frac{4}{5}x^6 - \frac{2009}{100}x^5 + \frac{2009}{25}x^4 + \frac{2251}{250}x^3 - \frac{4502}{125}x^2 - \frac{2}{5}x + \frac{8}{5} = 0 ,$$

Its roots are: 1) $x^2 - \frac{2}{5} = 0, x_1 = \sqrt{\frac{2}{5}}; x_2 = -\sqrt{\frac{2}{5}}$; 2) $x^2 - 100 = 0, x_3 = 10; x_4 = -10$;

3) $x^2 - \frac{1}{20} = 0, x_5 = \sqrt{\frac{1}{20}}; x_6 = -\sqrt{\frac{1}{20}}$; 4) $x - 4 = 0, x_7 = 4$.

… … ∞ … …

So that when $a \in R; a \neq 0; b = -4a; c = -\left(\frac{9a^2}{4} + \frac{4}{a}\right); d = 9a^2 + \frac{16}{a};$

$$e = \frac{a^3}{2} + 9; f = -\left(2a^3 + 36\right); g = -2a; h = 8a ;$$

the equation $ax^7 + bx^6 + cx^5 + dx^4 + ex^3 + fx^2 + gx + h = 0$, which has the roots: $x_1 = \sqrt{2a}; x_2 = -\sqrt{2a}$;

$x_3 = \frac{2}{a}; x_4 = -\frac{2}{a}; x_5 = \frac{\sqrt{a}}{2}; x_6 = -\frac{\sqrt{a}}{2}; x_7 = 4$. I call it one of the Dinbakish Theorems.

①: According to the equation $ax^7 + bx^6 + cx^5 + dx^4 + ex^3 + fx^2 + gx + h = 0$, can you fill in the blanks with the correct numbers for it?

[1]: ( ) + (-1) + ( ) + ( ) + (245) + ( ) + ( ) + (36) = 0 ;

[2]: ( ) + ( ) + (70) + ( ) + ( ) + ( ) + (180) + (36) = 0 ;

[3]: (640) + ( ) + (-2240) + ( ) + ( ) + ( ) + ( ) + (36) = 0 ;

[4]: ( ) + (-64) + ( ) + ( ) + ( ) + (-196) + ( ) + (36) = 0 ;

[5]: ( ) + ( ) + ( ) + (1134) + ( ) + ( ) + (540) + (36) = 0 ;

[6]: (-10935) + ( ) + ( ) + ( ) + (-6615) + ( ) + ( ) + (36) = 0 ;

87

$[7]$: $( )+\left(-\dfrac{1}{15625}\right)+\left(-\dfrac{14}{625}\right)+( )+\left(\dfrac{49}{25}\right)+( )+( )+(36)=0$;

$a=( );b=( );c=( );d=( );e=( );f=( );g=( )$;

$x_1=( );x_2=( );x_3=( );x_4=( );x_5=( );x_6=( );x_7=( )$.

ii: One of the forms of this equation can be:

$$\left(ax^2+6x-1\right)\left(x^2-4a\right)\left(x^2-\dfrac{a}{16}\right)\left(x-\dfrac{4}{a}\right)=0 ,$$

It expands:

$$ax^7+2x^6-\left(\dfrac{65a^2}{16}+\dfrac{24}{a}+1\right)x^5+\left(\dfrac{4}{a}-\dfrac{a}{8}-8a\right)x^4+\left(\dfrac{65a}{16}+\dfrac{195}{2}+\dfrac{a^3}{4}\right)x^3+\left(\dfrac{a^2}{2}-\dfrac{65}{4}\right)x^2-\left(\dfrac{a^2}{4}+6a\right)x+a=0 ,$$

And $b=2;c=-\left(\dfrac{65a^2}{16}+\dfrac{24}{a}+1\right);d=\dfrac{4}{a}-\dfrac{a}{8}-8a;e=\dfrac{65a}{16}+\dfrac{195}{2}+\dfrac{a^3}{4};f=\dfrac{a^2}{2}-\dfrac{65}{4};g=-\left(\dfrac{a^2}{4}+6a\right);h=a$.

The equation follows the real number a to be changed.

❶: when $a=4$, it means that:

$$c=-72;d=-\dfrac{63}{2};e=\dfrac{519}{4};f=-\dfrac{33}{4};g=-28;h=4 .$$

Now, the equation is:

$$4x^7+2x^6-72x^5-\dfrac{63}{2}x^4+\dfrac{519}{4}x^3-\dfrac{33}{4}x^2-28x+4=0 ,$$

Its roots are: 1) $4x^2+6x-1=0, x_1=\dfrac{-3+\sqrt{13}}{4};x_2=\dfrac{-3-\sqrt{13}}{4}$; 2) $x^2-16=0,x_3=4;x_4=-4$;

3) $x^2-\dfrac{1}{4}=0,x_5=\dfrac{1}{2};x_6=-\dfrac{1}{2}$; 4) $x-1=0,x_7=1$.

❷: when $a=8$, it means that:

$$c = -264; d = -\frac{129}{2}; e = 258; f = \frac{63}{4}; g = -64; h = 8 \ .$$

Now, the equation is:

$$8x^7 + 2x^6 - 264x^5 - \frac{129}{2}x^4 + 258x^3 - \frac{63}{4}x^2 - 64x + 8 = 0 \ ,$$

Its roots are: 1) $8x^2 + 6x - 1 = 0, x_1 = \frac{-3 + \sqrt{17}}{8}; x_2 = \frac{-3 - \sqrt{17}}{8}$; 2) $x^2 - 32 = 0, x_3 = 4\sqrt{2}; x_4 = -4\sqrt{2}$;

3) $x^2 - \frac{1}{2} = 0, x_5 = \frac{\sqrt{2}}{2}; x_6 = -\frac{\sqrt{2}}{2}$; 4) $x - \frac{1}{2} = 0, x_7 = \frac{1}{2}$.

… … ∞ … …

So that when $a \in R; a \neq 0; b = 2; c = -\left(\frac{65a^2}{16} + \frac{24}{a} + 1\right); d = \frac{4}{a} - \frac{a}{8} - 8a;$

$$e = \frac{65a}{16} + \frac{195}{2} + \frac{a^3}{4}; f = \frac{a^2}{2} - \frac{65}{4}; g = -\left(\frac{a^2}{4} + 6a\right); h = a \ ;$$

the equation $ax^7 + bx^6 + cx^5 + dx^4 + ex^3 + fx^2 + gx + h = 0$, which has the roots: $x_1 = \frac{-3 + \sqrt{9+a}}{a}$;

$x_2 = \frac{-3 - \sqrt{9+a}}{a}$; $x_3 = \sqrt{4a}; x_4 = -\sqrt{4a}; x_5 = \frac{\sqrt{a}}{4}; x_6 = -\frac{\sqrt{a}}{4}; x_7 = \frac{4}{a}$;

I call it one of the Dinbakish Theorems.

②: According to the equation $ax^7 + bx^6 + cx^5 + dx^4 + ex^3 + fx^2 + gx + h = 0$, can you fill in the blanks with the correct numbers for it?

[1]: $(\ ) + (\ ) + (-90) + (\ ) + (387) + (\ ) + (\ ) + (100) = 0$;

[2]: $(\ ) + (\ ) + (\ ) + (30) + (\ ) + (-129) + (\ ) + (100) = 0$;

[3]: $(384) + (-64) + (\ ) + (\ ) + (\ ) + (\ ) + (\ ) + (100) = 0$;

[4]: $(\ ) + (\ ) + (\ ) + (\ ) + (-3096) + (\ ) + (600) + (100) = 0$;

[5]: $(\ ) + (-15625) + (-281250) + (\ ) + (\ ) + (\ ) + (\ ) + (100) = 0$;

$[6]$: $(-234375)+(\ )+(\ )+(\ )+(\ )+(-48375)+(\ )+(100)=0$;

$[7]$: $(\ )+\left(-\dfrac{1}{729}\right)+(\ )+\left(\dfrac{10}{27}\right)+(\ )+\left(-\dfrac{43}{3}\right)+(\ )+(100)=0$;

$a=(\ );b=(\ );c=(\ );d=(\ );e=(\ );f=(\ );g=(\ )$;

$x_1=(\ );x_2=(\ );x_3=(\ );x_4=(\ );x_5=(\ );x_6=(\ );x_7=(\ )$.

iii: One of the forms of this equation can be:

$$\left(ax^2-7x+a\right)\left(x^2+4ax-2\right)\left(x^2-\dfrac{3a}{4}\right)\left(x+\dfrac{2}{a}\right)=0 \ ,$$

It expands:

$$ax^7+\left(4a^2-5\right)x^6-\left(\dfrac{3a^2}{4}+21a+\dfrac{14}{a}\right)x^5-\left(3a^3-\dfrac{15a}{4}+44\right)x^4+\left(\dfrac{63a^2}{4}+6a+\dfrac{28}{a}+\dfrac{21}{2}\right)x^3+\left(33a-3a^3-4\right)x^2-\left(\dfrac{9a^2}{2}+21\right)x+3a=0 \ ,$$

And

$$b=4a^2-5;c=-\left(\dfrac{3a^2}{4}+21a+\dfrac{14}{a}\right);d=-\left(3a^3-\dfrac{15a}{4}+44\right);e=\dfrac{63a^2}{4}+6a+\dfrac{28}{a}+\dfrac{21}{2};f=33a-3a^3-4;g=-\left(\dfrac{9a^2}{2}+21\right);h=3a \ .$$

The equation follows the real number a to be changed.

1: when $a=2$, it means that:

$$b=11;c=-52;d=-\dfrac{89}{2};e=\dfrac{199}{2};f=38;g=-39;h=6 \ .$$

Now, the equation is:

$$2x^7+11x^6-52x^5-\dfrac{89}{2}x^4+\dfrac{199}{2}x^3+38x^2-39x+6=0 \ ,$$

Its roots are: 1) $2x^2-7x+2=0, x_1=\dfrac{7+\sqrt{33}}{4};x_2=\dfrac{7-\sqrt{33}}{4}$; 2) $x^2+8x-2=0$,

$x_3 = -4 + \sqrt{18}; x_4 = -4 - \sqrt{18}$; 3) $x^2 - \dfrac{3}{2} = 0, x_5 = \sqrt{\dfrac{3}{2}}; x_6 = -\sqrt{\dfrac{3}{2}}$; 4) $x + 1 = 0, x_7 = -1$.

❷: when $a = -2$, it means that:

$b = 11; c = 46; d = -\dfrac{23}{2}; e = \dfrac{95}{2}; f = -46; g = -39; h = -6$.

Now, the equation is:

$-2x^7 + 11x^6 + 46x^5 - \dfrac{23}{2}x^4 + \dfrac{95}{2}x^3 - 46x^2 - 39x - 6 = 0$,

Its roots are: 1) $-2x^2 - 7x - 2 = 0, x_1 = \dfrac{-7 + \sqrt{33}}{4}; x_2 = \dfrac{-7 - \sqrt{33}}{4}$; 2) $x^2 - 8x - 2 = 0,$

$x_3 = 4 + \sqrt{18}; x_4 = 4 - \sqrt{18}$; 3) $x^2 + \dfrac{3}{2} = 0, x_5 = \dfrac{\sqrt{6}i}{2}; x_6 = -\dfrac{\sqrt{6}i}{2}$; 4) $x - 1 = 0$, $x_7 = 1$.

… … ∞ … …

So that when $a \in R; a \neq 0; b = 4a^2 - 5; c = -\left(\dfrac{3a^2}{4} + 21a + \dfrac{14}{a}\right); d = -\left(3a^3 - \dfrac{15a}{4} + 44\right); e = \dfrac{63a^2}{4} + 6a + \dfrac{28}{a} + \dfrac{21}{2};$

$$f = 33a - 3a^3 - 4; g = -\left(\dfrac{9a^2}{2} + 21\right); h = 3a \ ;$$

the equation $ax^7 + bx^6 + cx^5 + dx^4 + ex^3 + fx^2 + gx + h = 0$, which has the roots: $x_1 = -\dfrac{2}{a}$;

$x_2 = \dfrac{7 + \sqrt{49 - 4a^2}}{2a}; x_3 = \dfrac{7 - \sqrt{49 - 4a^2}}{2a}$; $x_4 = -2a + \sqrt{4a^2 + 2}; x_5 = -2a - \sqrt{4a^2 + 2}$;

$x_6 = \dfrac{\sqrt{3a}}{2}; x_7 = -\dfrac{\sqrt{3a}}{2}$, I call it one of the Dinbakish Theorems.

③: According to the equation $ax^7 + bx^6 + cx^5 + dx^4 + ex^3 + fx^2 + gx + h = 0$, can you fill in the blanks with the correct numbers for it?

[1]: ( )+( )+(−42)+( )+(168)+( )+( )+(64)=0 ;

[2]: $( )+(-1)+( )+(21)+( )+( )+( )+(64)=0$;

[3]: $(256)+( )+( )+( )+( )+( )+(-256)+(64)=0$;

[4]: $( )+( )+( )+(336)+( )+(-336)+( )+(64)=0$;

[5]: $( )+(-4096)+( )+(5376)+( )+( )+( )+(64)=0$;

[6]: $(-32768)+( )+( )+( )+(-10752)+( )+( )+(64)=0$;

[7]: $( )+\left(-\dfrac{1}{64}\right)+( )+\left(\dfrac{21}{16}\right)+( )+(-21)+( )+(64)=0$;

$a=( );b=( );c=( );d=( );e=( );f=( );g=( )$;

$x_1=( );x_2=( );x_3=( );x_4=( );x_5=( );x_6=( );x_7=( )$.

iv: One of the forms of this equation can be:

$$\left(x^2-3ax+3\right)\left(x^2+2x+a\right)\left(x^2-ax+1\right)\left(ax-3\right)=0 ,$$

It expands:

$$ax^7-\left(4a^2-2a+3\right)x^6+\left(3a^3-7a^2+16a-6\right)x^5+\left(2a^3-15a^2+29a-12\right)x^4+\left(3a^4-14a^2+21a-24\right)x^3-\left(15a^3-30a+9\right)x^2+\left(21a^2-18\right)x-9a=0 ,$$

And $b=-\left(4a^2-2a+3\right);c=3a^3-7a^2+16a-6;d=2a^3-15a^2+29a-12;$

$e=3a^4-14a^2+21a-24;f=-\left(15a^3-30a+9\right);g=21a^2-18;h=-9a$.

The equation follows the real number a to be changed.

❶ : when $a=3$, it means that:

$b=-33;c=60;d=-6;e=156;f=-324;g=171;h=-27$.

Now, the equation is:

$$3x^7-33x^6+60x^5-6x^4+156x^3-324x^2+171x-27=0 ,$$

92

Its roots are: 1) $x^2 - 9x + 3 = 0, x_1 = \dfrac{9+\sqrt{69}}{2}; x_2 = \dfrac{9-\sqrt{69}}{2}$; 2) $x^2 + 2x + 3 = 0, x_3 = -1 + \sqrt{2}i$;

$x_4 = -1 - \sqrt{2}i$; 3) $x^2 - 3x + 1 = 0, x_5 = \dfrac{3+\sqrt{5}}{2}; x_6 = \dfrac{3-\sqrt{5}}{2}$; 4) $3x - 3 = 0, x_7 = 1$.

❷ : when $a = -3$, it means that:

$b = -45; c = -198; d = -288; e = 30; f = 306; g = 171; h = 27$.

Now, the equation is:

$-3x^7 - 45x^6 - 198x^5 - 288x^4 + 30x^3 + 306x^2 + 171x + 27 = 0$,

Its roots are: 1) $x^2 + 9x + 3 = 0, x_1 = \dfrac{-9+\sqrt{69}}{2}; x_2 = \dfrac{-9-\sqrt{69}}{2}$; 2) $x^2 + 2x - 3 = 0, x_3 = 1; x_4 = -3$;

3) $x^2 + 3x + 1 = 0, x_5 = \dfrac{-3+\sqrt{5}}{2}; x_6 = \dfrac{-3-\sqrt{5}}{2}$; 4) $-3x - 3 = 0, x_7 = -1$.

… … ∞ … …

So that when $a \in R; a \neq 0; b = -\left(4a^2 - 2a + 3\right); c = 3a^3 - 7a^2 + 16a - 6; d = 2a^3 - 15a^2 + 29a - 12;$

$e = 3a^4 - 14a^2 + 21a - 24; f = -\left(15a^3 - 30a + 9\right); g = 21a^2 - 18; h = -9a$.

The equation $ax^7 + bx^6 + cx^5 + dx^4 + ex^3 + fx^2 + gx + h = 0$, which has the roots: $x_1 = \dfrac{3}{a}$;

$x_2 = \dfrac{3a+\sqrt{9a^2-3}}{2}; x_3 = \dfrac{3a-\sqrt{9a^2-3}}{2}$; $x_4 = -1+\sqrt{1-a}; x_5 = -1-\sqrt{1-a}$;

$x_6 = \dfrac{a+\sqrt{a^2-4}}{2}; x_7 = \dfrac{a-\sqrt{a^2-4}}{2}$. I call it one of the Dinbakish Theorems.

④: According to the equation $ax^7 + bx^6 + cx^5 + dx^4 + ex^3 + fx^2 + gx + h = 0$, can you fill in the blanks with the correct numbers for it?

[1]: ( ) + (−1) + ( ) + (9) + ( ) + (−26) + ( ) + (24) = 0 ;

[2]: $(\ )+(-27)+(\ )+(81)+(\ )+(\ )+(\ )+(24)=0$;

[3]: $(\ )+(\ )+(\ )+(\ )+(\ )+(-78)+(\ )+(24)=0$;

[4]: $(\ )+(\ )+(\ )+(36)+(\ )+(\ )+(\ )+(24)=0$;

[5]: $(\ )+(-8)+(\ )+(\ )+(\ )+(-52)+(\ )+(24)=0$;

[6]: $(128)+(\ )+(-288)+(\ )+(\ )+(\ )+(\ )+(24)=0$;

[7]: $(\ )+(\ )+(\ )+(\ )+(-208)+(\ )+(48)+(24)=0$;

$a=(\ ); b=(\ ); c=(\ ); d=(\ ); e=(\ ); f=(\ ); g=(\ )$;

$x_1=(\ ); x_2=(\ ); x_3=(\ ); x_4=(\ ); x_5=(\ ); x_6=(\ ); x_7=(\ )$.

(4) : An Equation

$$ax^8 + bx^7 + cx^6 + dx^5 + ex^4 + fx^3 + gx^2 + hx + i = 0$$

$$(\ a \in R; b \in R; c \in R; d \in R; e \in R; f \in R; g \in R; h \in R; i \in R\ )$$

$$(\ a \neq 0; b \neq 0; c \neq 0; d \neq 0; e \neq 0; f \neq 0; g \neq 0; h \neq 0; i \neq 0\ )$$

i: One of the forms of this equation can be:

$$\left(x^2 - a\right)\left(x^2 - \frac{a}{5}\right)\left(x^2 - \frac{25}{a^2}\right)\left(x - \frac{a}{5}\right)(ax+10) = 0 ,$$

It expands:

$$ax^8 + \left(10 - \frac{a^2}{5}\right)x^7 - \left(\frac{6a^2}{5} + \frac{25}{a} + 2a\right)x^6 - \left(10 - \frac{a^2}{5}\right)\left(\frac{6a}{5} + \frac{25}{a^2}\right)x^5 + \left(\frac{a^3}{5} + \frac{12a^2}{5} + \frac{50}{a} + 30\right)x^4 + \left(10 - \frac{a^2}{5}\right)\left(\frac{30}{a} + \frac{a^2}{5}\right)x^3 - \left(\frac{2a^3}{5} + 5a + 60\right)x^2 - \left(50 - a^2\right)x + 10a = 0 ,$$

And $b = 10 - \frac{a^2}{5}$; $c = -\left(\frac{6a^2}{5} + \frac{25}{a} + 2a\right)$; $d = -\left(10 - \frac{a^2}{5}\right)\left(\frac{6a}{5} + \frac{25}{a^2}\right)$; $e = \frac{a^3}{5} + \frac{12a^2}{5} + \frac{50}{a} + 30$;

94

$$f = \left(10 - \frac{a^2}{5}\right)\left(\frac{30}{a} + \frac{a^2}{5}\right); g = -\left(\frac{2a^3}{5} + 5a + 60\right); h = -\left(50 - a^2\right); i = 10a \ .$$

The equation follows the real number a to be changed.

1 : when $a = 5$, it means that:

$$b = -5; c = -45; d = -35; e = 125; f = 55; g = -135; h = -25; i = 50 \ .$$

Now, the equation is:

$$5x^8 - 5x^7 - 45x^6 - 35x^5 + 125x^4 + 55x^3 - 135x^2 - 25x + 50 = 0 \ ,$$

Its roots are: 1) $x^2 - 5 = 0, x_1 = \sqrt{5}; x_2 = -\sqrt{5}$; 2) $x^2 - 1 = 0, x_3 = 1; x_4 = -1$; 3) $x^2 - 1 = 0$, $x_5 = 1$;

$$x_6 = -1 \ ; \ 4) \ 5x + 10 = 0, x_7 = -2 \ .$$

2 : when $a = 10$, it means that:

$$b = -10; c = -\frac{285}{2}; d = \frac{245}{2}; e = 475; f = -230; g = -510; h = 50; i = 100 \ .$$

Now, the equation is:

$$10x^8 - 10x^7 - \frac{285}{2}x^6 + \frac{245}{2}x^5 + 475x^4 - 230x^3 - 510x^2 + 50x + 100 = 0 \ ,$$

Its roots are: 1) $x^2 - 10 = 0, x_1 = \sqrt{10}; x_2 = -\sqrt{10}$; 2) $x^2 - 2 = 0, x_3 = \sqrt{2}; x_4 = -\sqrt{2}$;

3) $x^2 - \frac{1}{4} = 0, x_5 = \frac{1}{2}; x_6 = -\frac{1}{2}$; 4) $x - 2 = 0, x_7 = 2$; 5) $10x + 10 = 0, x_8 = -1$.

… … ∞ … …

So that when $a \in R; a \neq 0; b = 10 - \frac{a^2}{5}; c = -\left(\frac{6a^2}{5} + \frac{25}{a} + 2a\right); d = -\left(10 - \frac{a^2}{5}\right)\left(\frac{6a}{5} + \frac{25}{a^2}\right)$;

$e = \frac{a^3}{5} + \frac{12a^2}{5} + \frac{50}{a} + 30$; $f = \left(10 - \frac{a^2}{5}\right)\left(\frac{30}{a} + \frac{a^2}{5}\right); g = -\left(\frac{2a^3}{5} + 5a + 60\right); h = -\left(50 - a^2\right); i = 10a$;

the equation $ax^8 + bx^7 + cx^6 + dx^5 + ex^4 + fx^3 + gx^2 + hx + i = 0$, which has the roots: $x_1 = \sqrt{a}$;

$$x_2 = -\sqrt{a}; x_3 = \sqrt{\frac{a}{5}}; x_4 = -\sqrt{\frac{a}{5}}; x_5 = \frac{5}{a}; x_6 = -\frac{5}{a}; x_7 = \frac{a}{5}; x_8 = -\frac{10}{a} \ .$$

I call it one of the Dinbakish Theorems.

①: According to the equation $ax^8 + bx^7 + cx^6 + dx^5 + ex^4 + fx^3 + gx^2 + hx + i = 0$, can you fill in the blanks with the correct numbers for it?

$[1]$: $(\ )+(-8)+(\ )+(56)+(\ )+(\ )+(\ )+(\ )+(-120) = 0$;

$[2]$: $(\ )+(\ )+(8)+(\ )+(\ )+(112)+(\ )+(-64)+(-120) = 0$;

$[3]$: $(16)+(\ )+(\ )+(\ )+(-364)+(\ )+(\ )+(\ )+(-120) = 0$;

$[4]$: $(\ )+(\ )+(64)+(\ )+(\ )+(\ )+(404)+(\ )+(-120) = 0$;

$[5]$: $(\ )+(-1024)+(\ )+(1792)+(\ )+(-896)+(\ )+(\ )+(-120) = 0$;

$[6]$: $(256)+(\ )+(\ )+(\ )+(\ )+(\ )+(\ )+(-128)+(-120) = 0$;

$[7]$: $(\ )+(\ )+(5832)+(13608)+(\ )+(\ )+(\ )+(\ )+(-120) = 0$;

$[8]$: $(\ )+(-625000)+(\ )+(\ )+(-56875)+(-14000)+(\ )+(\ )+(-120) = 0$;

$a = (\ ); b = (\ ); c = (\ ); d = (\ ); e = (\ ); f = (\ ); g = (\ ); h = (\ )$;

$x_1 = (\ ); x_2 = (\ ); x_3 = (\ ); x_4 = (\ ); x_5 = (\ ); x_6 = (\ ); x_7 = (\ ); x_8 = (\ )$.

ii: One of the forms of this equation can be:

$$\left(ax^2 + 6x + a\right)\left(x^2 - \frac{6}{a^2}\right)\left(x^2 - \frac{a}{6}\right)\left(x^2 - 2a\right) = 0$$

It expands:

$$ax^8 + 6x^7 - \left(\frac{6}{a} + \frac{13a^2}{6} - a\right)x^6 - \left(\frac{36}{a^2} + 13a\right)x^5 - \left(\frac{6}{a} + \frac{13a^2}{6} - 13 - \frac{a^3}{3}\right)x^4 + \left(\frac{78}{a} + 2a^2\right)x^3 + \left(13 + \frac{a^3}{3} - 2a\right)x^2 - 12x - 2a = 0 \ ,$$

96

And $b = 6; c = -\left(\dfrac{6}{a} + \dfrac{13a^2}{6} - a\right); d = -\left(\dfrac{36}{a^2} + 13a\right); \quad e = -\left(\dfrac{6}{a} + \dfrac{13a^2}{6} - 13 - \dfrac{a^3}{3}\right);$

$$f = \dfrac{78}{a} + 2a^2; g = 13 + \dfrac{a^3}{3} - 2a; h = -12; i = -2a .$$

The equation follows the real number a to be changed.

❶ : when $a = 6$, it means that:

$$c = -73; d = -79; e = 6; f = 85; g = 73; i = -12 .$$

Now, the equation is:

$$6x^8 + 6x^7 - 73x^6 - 79x^5 + 6x^4 + 85x^3 + 73x^2 - 12x - 12 = 0 ,$$

Its roots are: 1) $6x^2 + 6x + 6 = 0, x_1 = \dfrac{-1 + \sqrt{3}i}{2}; x_2 = \dfrac{-1 - \sqrt{3}i}{2}$; 2) $x^2 - \dfrac{1}{6} = 0, x_3 = \dfrac{\sqrt{6}}{6}; x_4 = -\dfrac{\sqrt{6}}{6}$;

  3) $x^2 - 1 = 0, x_5 = 1; x_6 = -1$; 4) $x^2 - 12 = 0, x_7 = 2\sqrt{3}; x_8 = -2\sqrt{3}$.

❷ : when $a = -3$, it means that:

$$c = -\dfrac{41}{2}; d = 35; e = -\dfrac{27}{2}; f = -8; g = 10; i = 6 .$$

Now, the equation is:

$$-3x^8 + 6x^7 - \dfrac{41}{2}x^6 + 35x^5 - \dfrac{27}{2}x^4 - 8x^3 + 10x^2 - 12x + 6 = 0 ,$$

Its roots are: 1) $-3x^2 + 6x - 3 = 0, x_1 = x_2 = 1$; 2) $x^2 - \dfrac{2}{3} = 0, x_3 = \dfrac{\sqrt{6}}{3}; x_4 = -\dfrac{\sqrt{6}}{3}$;

  3) $x^2 + \dfrac{1}{2} = 0, x_5 = \dfrac{\sqrt{2}i}{2}; x_6 = -\dfrac{\sqrt{2}i}{2}$; 4) $x^2 + 6 = 0, x_7 = \sqrt{6}i; x_8 = -\sqrt{6}i$.

… … ∞ … …

So that when $a \in R; a \neq 0; b = 6; c = -\left(\dfrac{6}{a} + \dfrac{13a^2}{6} - a\right); d = -\left(\dfrac{36}{a^2} + 13a\right); e = -\left(\dfrac{6}{a} + \dfrac{13a^2}{6} - 13 - \dfrac{a^3}{3}\right);$

$f = \dfrac{78}{a} + 2a^2; g = 13 + \dfrac{a^3}{3} - 2a; h = -12; i = -2a$, the equation $ax^8 + bx^7 + cx^6 + dx^5 + ex^4 + fx^3 + gx^2 + hx + i = 0$

, which has the roots: $x_1 = \dfrac{\sqrt{6}}{a}; x_2 = -\dfrac{\sqrt{6}}{a}; x_3 = \dfrac{\sqrt{a}}{6}; x_4 = -\dfrac{\sqrt{a}}{6}; x_5 = \sqrt{2a}; x_6 = -\sqrt{2a}$;

$x_7 = \dfrac{-3 + \sqrt{9 - a^2}}{a}; x_8 = \dfrac{-3 - \sqrt{9 - a^2}}{a}$. I call it one of the Dinbakish Theorems.

②: According to the equation $ax^8 + bx^7 + cx^6 + dx^5 + ex^4 + fx^3 + gx^2 + hx + i = 0$, can you fill in the blanks with the correct numbers for it?

$[1]$: $(\ )+(2)+(\ )+(\ )+(\ )+(52)+(\ )+(\ )+(72)=0$;

$[2]$: $(\ )+(\ )+(\ )+(4374)+(\ )+(\ )+(\ )+(144)+(72)=0$;

$[3]$: $(\ )+(\ )+(\ )+(\ )+(848)+(\ )+(\ )+(96)+(72)=0$;

$[4]$: $(\ )+(-256)+(\ )+(\ )+(\ )+(416)+(\ )+(\ )+(72)=0$;

$[5]$: $(16)+(\ )+(\ )+(\ )+(212)+(\ )+(\ )+(\ )+(72)=0$;

$[6]$: $(\ )+(\ )+(-96)+(\ )+(\ )+(\ )+(-204)+(\ )+(72)=0$;

$[7]$: $(81)+(\ )+(\ )+(\ )+(477)+(\ )+(\ )+(\ )+(72)=0$;

$[8]$: $(\ )+(\ )+(-324)+(\ )+(\ )+(\ )+(-306)+(\ )+(72)=0$;

$a = (\ ); b = (\ ); c = (\ ); d = (\ ); e = (\ ); f = (\ ); g = (\ ); h = (\ )$;

$x_1 = (\ ); x_2 = (\ ); x_3 = (\ ); x_4 = (\ ); x_5 = (\ ); x_6 = (\ ); x_7 = (\ ); x_8 = (\ )$.

iii: One of the forms of this equation can be:

$$\left(ax^2 - x + a\right)\left(x^2 + ax - 1\right)\left(x^2 - \frac{4}{a}\right)\left(x^2 - 4a\right) = 0 ,$$

It expands:

$$ax^8 + \left(a^2 - 1\right)x^7 - \left(4a^2 + a + 4\right)x^6 + \left[a^2 + 1 - \left(a^2 - 1\right)\left(4a + \frac{4}{a}\right)\right]x^5 + \left(4a^2 + 4 + 15a\right)x^4 + \left[16a^2 - 16 - \left(a^2 + 1\right)\left(4a + \frac{4}{a}\right)\right]x^3 + \left(4a^2 + 4 - 16a\right)x^2 + \left(16a^2 + 16\right)x - 16a = 0 ,$$

And $b = a^2 - 1; c = -\left(4a^2 + a + 4\right); d = a^2 + 1 - \left(a^2 - 1\right)\left(4a + \frac{4}{a}\right); e = 4a^2 + 4 + 15a$;

$$f = 16a^2 - 16 - \left(a^2 + 1\right)\left(4a + \frac{4}{a}\right); g = 4a^2 + 4 - 16a; h = 16a^2 + 16; i = -16a .$$

The equation follows the real number a to be changed.

1 : when $a = 4$, it means that;

$b = 15; c = -72; d = -238; e = 128; f = -49; g = 4; h = 272; i = -64$.

Now, the equation is:

$$4x^8 + 15x^7 - 72x^6 - 238x^5 + 128x^4 - 49x^3 + 4x^2 + 272x - 64 = 0 ,$$

Its roots are: 1) $4x^2 - x + 4 = 0, x_1 = \dfrac{1 + \sqrt{63}i}{8}; x_2 = \dfrac{1 - \sqrt{63}i}{8}$; 2) $x^2 + 4x - 1 = 0, x_3 = -2 + \sqrt{5}$;

$x_4 = -2 - \sqrt{5}$; 3) $x^2 - 1 = 0, x_5 = 1; x_6 - 1$; 4) $x^2 - 16 = 0, x_7 = 4; x_8 = -4$.

2 : when $a = \dfrac{1}{4}$, it means that:

$$b = -\frac{15}{16}; c = -\frac{9}{2}; d = \frac{272}{16}; e = 8; f = -\frac{529}{16}; g = \frac{1}{4}; h - 17; i - -4 .$$

Now, the equation is:

$$\frac{1}{4}x^8 - \frac{15}{16}x^7 - \frac{9}{2}x^6 + \frac{272}{16}x^5 + 8x^4 - \frac{529}{16}x^3 + \frac{1}{4}x^2 + 17x - 4 = 0 ,$$

Its roots are: 1) $\dfrac{1}{4}x^2 - x + \dfrac{1}{4} = 0, x_1 = 2 + \sqrt{3}; x_2 = 2 - \sqrt{3}$; 2) $x^2 + \dfrac{1}{4}x - 1 = 0, x_3 = \dfrac{-1 + \sqrt{65}}{8}$;

$x_4 = \dfrac{-1-\sqrt{65}}{8}$; 3) $x^2 - 16 = 0, x_5 = 4; x_6 = -4$; 4) $x^2 - 1 = 0, x_7 = 1; x_8 = -1$.

… … ∞ … …

So that when $a \in R; a \neq 0; b = a^2 - 1; c = -\left(4a^2 + a + 4\right); d = a^2 + 1 - \left(a^2 - 1\right)\left(4a + \dfrac{4}{a}\right); e = 4a^2 + 4 + 15a$;

$f = 16a^2 - 16 - \left(a^2 + 1\right)\left(4a + \dfrac{4}{a}\right); g = 4a^2 + 4 - 16a; h = 16a^2 + 16; i = -16a$, the equation

$ax^8 + bx^7 + cx^6 + dx^5 + ex^4 + fx^3 + gx^2 + hx + i = 0$, which has the roots: $x_1 = 2\sqrt{a}; x_2 = -2\sqrt{a}$;

$x_3 = \sqrt{\dfrac{4}{a}}; x_4 = -\sqrt{\dfrac{4}{a}}$; $x_5 = \dfrac{1 + \sqrt{1 - 4a^2}}{2a}; x_6 = \dfrac{1 - \sqrt{1 - 4a^2}}{2a}$; $x_7 = \dfrac{-a + \sqrt{a^2 + 4}}{2}; x_8 = \dfrac{-a - \sqrt{a^2 + 4}}{2}$.

I call it one of the Dinbakish Theorems.

③: According to the equation $ax^8 + bx^7 + cx^6 + dx^5 + ex^4 + fx^3 + gx^2 + hx + i = 0$, can you fill in the blanks with the correct numbers for it?

[1]: ( )+(1)+( )+(-10)+( )+(31)+( )+( )+(60)=0 ;

[2]: ( )+(-128)+( )+(320)+( )+(-248)+( )+( )+(60)=0 ;

[3]: (16)+( )+( )+( )+( )+( )+(-184)+( )+(60)=0 ;

[4]: ( )+( )+(-96)+( )+(204)+( )+( )+( )+(60)=0 ;

[5]: (81)+( )+( )+( )+(459)+( )+( )+( )+(60)=0 ;

[6]: ( )+( )+(-324)+( )+( )+( )+(-276)+( )+(60)=0 ;

[7]: ( )+( )+( )+( )+(1275)+( )+(-460)+( )+(60)=0 ;

[8]: (625)+( )+(-1500)+( )+( )+( )+( )+( )+(60)=0 ;

$a = (\ ); b = (\ ); c = (\ ); d = (\ ); e = (\ ); f = (\ ); g = (\ ); h = (\ )$;

$x_1 = (\ ); x_2 = (\ ); x_3 = (\ ); x_4 = (\ ); x_5 = (\ ); x_6 = (\ ); x_7 = (\ ); x_8 = (\ )$.

iv: One of the forms of this equation can be:

$$(ax^2 + 2x + 3)(x^2 + 3ax - 2)(x^2 + x + a)(x^2 + a) = 0 ,$$

It expands:

$$ax^8 + (3a^2 + a + 2)x^7 + (5a^2 + 4a + 5)x^6 + (6a^3 + a^2 + 17a - 1)x^5 + (4a^3 + 8a^2 + 17a - 10)x^4 + (3a^4 + 24a^2 - 5a - 6)x^3 + (4a^3 + 12a^2 - 16a)x^2 + (9a^3 - 4a^2 - 6a)x - 6a^2 = 0 ,$$

And $b = 3a^2 + a + 2; c = 5a^2 + 4a + 5; d = 6a^3 + a^2 + 17a - 1; e = 4a^3 + 8a^2 + 17a - 10$;

$$f = 3a^4 + 24a^2 - 5a - 6; g = 4a^3 + 12a^2 - 16a; h = 9a^3 - 4a^2 - 6a; i = -6a^2 .$$

The equation follows the real number a to be changed.

①: when $a = -1$; it means that:

$b = 4; c = 6; d = -23; e = -23; f = 26; g = 24; h = -7; i = -6$.

Now, the equation is:

$$-x^8 + 4x^7 + 6x^6 - 23x^5 - 23x^4 + 26x^3 + 24x^2 - 7x - 6 = 0 ,$$

Its roots are: 1) $-x^2 + 2x + 3 = 0, x_1 = 1; x_2 = -3$; 2) $x^2 - 3x - 2 = 0, x_3 = \dfrac{3 + \sqrt{17}}{2}; x_4 = \dfrac{3 - \sqrt{17}}{2}$;

3) $x^2 + x - 1 = 0, x_5 = \dfrac{-1 + \sqrt{5}}{2}; x_6 = \dfrac{-1 - \sqrt{5}}{2}$; 4) $x^2 - 1 = 0, x_7 = 1; x_8 = -1$.

②: when $a = -4$, it means that:

$b = 46; c = 69; d - 437; e = -206; f = 1166; g = 0; h = -616; i = -96$.

Now, the equation is:

$$-4x^8 + 46x^7 + 69x^6 - 437x^5 - 206x^4 + 1166x^3 - 616x - 96 = 0 ,$$

Its roots are: 1) $-4x^2 + 2x + 3 = 0, x_1 = \dfrac{1 + \sqrt{13}}{4}; x_2 = \dfrac{1 - \sqrt{13}}{4}$; 2) $x^2 - 12x - 2 = 0, x_3 = 6 + \sqrt{38};$

$x_4 = 6 - \sqrt{38}$; 3) $x^2 + x - 4 = 0, x_5 = \dfrac{-1+\sqrt{17}}{2}; x_6 = \dfrac{-1-\sqrt{17}}{2}$; 4) $x^2 - 4 = 0, x_7 = 2; x_8 = -2$.

… … ∞ … …

So that when $a \in R; a \neq 0; b = 3a^2 + a + 2; c = 5a^2 + 4a + 5; d = 6a^3 + a^2 + 17a - 1; e = 4a^3 + 8a^2 + 17a - 10$;
$f = 3a^4 + 24a^2 - 5a - 6; g = 4a^3 + 12a^2 - 16a; h = 9a^3 - 4a^2 - 6a; i = -6a^2$, the equation
$ax^8 + bx^7 + cx^6 + dx^5 + ex^4 + fx^3 + gx^2 + hx + i = 0$, which has the roots: $x_1 = \sqrt{-a}; x_2 = -\sqrt{a}$;

$$x_3 = \frac{-1+\sqrt{1-3a}}{a}; x_4 = \frac{-1-\sqrt{1-3a}}{a} \ , \ x_5 = \frac{-3a+\sqrt{9a^2+8}}{2}; x_6 = \frac{-3a-\sqrt{9a^2+8}}{2};$$

$$x_7 = \frac{-1+\sqrt{1-4a}}{2}; x_8 = \frac{-1-\sqrt{1-4a}}{2}; \ \text{I call it one of the Dinbakish Theorems.}$$

④: According to the equation $ax^8 + bx^7 + cx^6 + dx^5 + ex^4 + fx^3 + gx^2 + hx + i = 0$, can you fill in the blanks with the correct numbers for it?

[1]: $(81) + (\ \ ) + (-540) + (\ \ ) + (\ \ ) + (\ \ ) + (\ \ ) + (\ \ ) + (144) = 0$;

[2]: $(\ \ ) + (\ \ ) + (\ \ ) + (\ \ ) + (1035) + (\ \ ) + (-720) + (\ \ ) + (144) = 0$;

[3]: $(\ \ ) + (1) + (\ \ ) + (-8) + (\ \ ) + (\ \ ) + (\ \ ) + (\ \ ) + (144) = 0$;

[4]: $(\ \ ) + (\ \ ) + (\ \ ) + (\ \ ) + (\ \ ) + (-19) + (\ \ ) + (12) + (144) = 0$;

[5]: $(6561) + (\ \ ) + (-14580) + (\ \ ) + (\ \ ) + (\ \ ) + (\ \ ) + (\ \ ) + (144) = 0$;

[6]: $(\ \ ) + (-16384) + (\ \ ) + (8192) + (\ \ ) + (\ \ ) + (\ \ ) + (\ \ ) + (144) = 0$;

[7]: $(\ \ ) + (\ \ ) + (\ \ ) + (\ \ ) + (1840) + (\ \ ) + (\ \ ) + (-24) + (144) = 0$;

[8]: $(\ \ ) + (\ \ ) + (\ \ ) + (\ \ ) + (\ \ ) + (-152) + (-960) + (\ \ ) + (144) = 0$;

$a = (\ \ ); b = (\ \ ); c = (\ \ ); d = (\ \ ); e = (\ \ ); f = (\ \ ); g = (\ \ ); h = (\ \ )$;

$x_1 = (\ \ ); x_2 = (\ \ ); x_3 = (\ \ ); x_4 = (\ \ ); x_5 = (\ \ ); x_6 = (\ \ ); x_7 = (\ \ ); x_8 = (\ \ )$.

One of the forms of this equation can be:

$$\left(ax^2 - 2x + a\right)\left(x^2 + ax - 1\right)\left(x^2 + x - a\right)\left(x^2 - ax - 2\right) = 0 ,$$

It expands:

$$ax^8 + (a-2)x^7 - \left(a^3 + a^2 + 2a + 2\right)x^6 - \left(a^3 - a^2 - 6\right)x^5 + \left(a^4 - a^3 + 3a^2 + a + 6\right)x^4 - \left(2a^3 + a^2 + 5a + 4\right)x^3 + \left(a^4 - 2a^2 + 2a - 4\right)x^2 + \left(a^3 + 6a\right)x - 2a^2 = 0 ,$$

And $b = a - 2; c = -\left(a^3 + a^2 + 2a + 2\right); d = -\left(a^3 - a^2 - 6\right); e = a^4 - a^3 + 3a^2 + a + 6$;

$f = -\left(2a^3 + a^2 + 5a + 4\right); g = a^4 - 2a^2 + 2a - 4; h = a^3 + 6a; i = -2a^2$.

The equation follows the real number a to be changed.

1 : when $a = 1$, it means that:

$b = -1; c = -6; d = 6; e = 10; f = -12; g = -3; h = 7; i = -2$.

Now, the equation is:

$$x^8 - x^7 - 6x^6 + 6x^5 + 10x^4 - 12x^3 - 3x^2 + 7x - 2 = 0 ,$$

Its roots are: 1) $x^2 - 2x + 1 = 0, x_1 = x_2 = 1$; 2) $x^2 + x - 1 = 0, x_3 = \dfrac{-1 + \sqrt{5}}{2}; x_4 = \dfrac{-1 - \sqrt{5}}{2}$;

3) $x^2 + x - 1 = 0, x_5 = \dfrac{-1 + \sqrt{5}}{2}; x_6 = \dfrac{-1 - \sqrt{5}}{2}$; 4) $x^2 - x - 2 = 0, x_7 = -1; x_8 = 2$.

2 : when $a = -1$, it means that:

$b = -3; c = 0; d = 8; e = 10; f = 2; g = -7; h = -7; i = -2$.

Now, the equation is:

$$-x^8 - 3x^7 + 8x^5 + 10x^4 + 2x^3 - 7x^2 - 7x - 2 = 0 ,$$

Its roots are: 1) $-x^2 - 2x - 1 = 0, x_1 = x_2 = -1$; 2) $x^2 - x - 1 = 0, x_3 = \dfrac{1 + \sqrt{5}}{2}; x_4 = \dfrac{1 - \sqrt{5}}{2}$;

3) $x^2 + x + 1 = 0, x_5 = \dfrac{-1+\sqrt{3}i}{2}; x_6 = \dfrac{-1-\sqrt{3}i}{2}$; 4) $x^2 + x - 2 = 0; x_7 = 1; x_8 = -2$.

… … ∞ … …

So that when $a \in R; a \neq 0; b = a - 2; c = -(a^3 + a^2 + 2a + 2); d = -(a^3 - a^2 - 6); e = a^4 - a^3 + 3a^2 + a + 6$;

$f = -(2a^3 + a^2 + 5a + 4); g = a^4 - 2a^2 + 2a - 4; h = a^3 + 6a; i = -2a^2$, the equation

$ax^8 + bx^7 + cx^6 + dx^5 + ex^4 + fx^3 + gx^2 + hx + i = 0$, which has the roots: $x_1 = \dfrac{1+\sqrt{1-a^2}}{a}$;

$x_2 = \dfrac{1-\sqrt{1-a^2}}{a}$; $x_3 = \dfrac{-a+\sqrt{a^2+4}}{2}; x_4 = \dfrac{-a-\sqrt{a^2+4}}{2}$; $x_5 = \dfrac{-1+\sqrt{1+4a}}{2}; x_6 = \dfrac{-1-\sqrt{1+4a}}{2}$;

$x_7 = \dfrac{a+\sqrt{a^2+8}}{2}; x_8 = \dfrac{a-\sqrt{a^2+8}}{2}$. I call it one of the Dinbakish Theorems.

⑤: According to the equation $ax^8 + bx^7 + cx^6 + dx^5 + ex^4 + fx^3 + gx^2 + hx + i = 0$, can you fill in the blanks with the correct numbers for it?

[1]: $(16) + (\ ) + (-152) + (\ ) + (\ ) + (\ ) + (\ ) + (\ ) + (96) = 0$;

[2]: $(\ ) + (\ ) + (\ ) + (\ ) + (392) + (\ ) + (-352) + (\ ) + (96) = 0$;

[3]: $(\ ) + (1) + (\ ) + (-7) + (\ ) + (\ ) + (\ ) + (\ ) + (96) = 0$;

[4]: $(\ ) + (\ ) + (\ ) + (\ ) + (\ ) + (-14) + (\ ) + (8) + (96) = 0$;

[5]: $(6561) + (\ ) + (-13851) + (\ ) + (\ ) + (\ ) + (\ ) + (\ ) + (96) = 0$;

[6]: $(\ ) + (-16384) + (\ ) + (7168) + (\ ) + (\ ) + (\ ) + (\ ) + (96) = 0$;

[7]: $(\ ) + (\ ) + (\ ) + (\ ) + (1568) + (\ ) + (\ ) + (-16) + (96) = 0$;

[8]: $(\ ) + (\ ) + (\ ) + (\ ) + (\ ) + (-112) + (-704) + (\ ) + (96) = 0$;

$a = (\ ); b = (\ ); c = (\ ); d = (\ ); e = (\ ); f = (\ ); g = (\ ); h = (\ );$

$x_1 = (\ ); x_2 = (\ ); x_3 = (\ ); x_4 = (\ ); x_5 = (\ ); x_6 = (\ ); x_7 = (\ ); x_8 = (\ )$.

$$(5) : \text{An Equation}$$

$$ax^9 + bx^8 + cx^7 + dx^6 + ex^5 + fx^4 + gx^3 + hx^2 + ix + j = 0$$

$$a \in R; b \in R; c \in R; d \in R; e \in R; f \in R; g \in R; h \in R; i \in R; j \in R;$$

$$a \neq 0; b \neq 0; c \neq 0; d \neq 0; e \neq 0; f \neq 0; g \neq 0; h \neq 0; i \neq 0; j \neq 0;$$

One of the forms of this equation can be:

$$\left(x^2 - \frac{a}{2}\right)\left(x^2 - \frac{3a}{2}\right)\left(x^2 - \frac{4}{a}\right)\left(x^2 - \frac{3}{a}\right)(ax + 1) = 0 ,$$

It expands:

$$ax^9 + x^8 - \left(7 + 2a^2\right)x^7 - \left(\frac{7}{a} + 2a\right)x^6 + \left(\frac{12}{a} + 14a + \frac{3a^3}{4}\right)x^5 + \left(\frac{12}{a^2} + 14 + \frac{3a^2}{4}\right)x^4 - \left(24 + \frac{21a^2}{4}\right)x^3 - \left(\frac{24}{a} + \frac{21a}{4}\right)x^2 + 9ax + 9 = 0 ,$$

And $b = 1; c = -\left(7 + 2a^2\right); d = -\left(\frac{7}{a} + 2a\right); e = \frac{12}{a} + 14a + \frac{3a^3}{4}; f = \frac{12}{a^2} + 14 + \frac{3a^2}{4}$;

$$g = -\left(24 + \frac{21a^2}{4}\right); h = -\left(\frac{24}{a} + \frac{21a}{4}\right); i = 9a; j = 9 .$$

The equation follows the real number a to be changed.

1 : when $a = -2$, it means that:

$$c = -15; d = \frac{15}{2}; e = -40; f = 20; g = -45; h = \frac{45}{2}; i = -18 .$$

Now, the equation is:

$$-2x^9 + x^8 - 15x^7 + \frac{15}{2}x^6 - 40x^5 + 20x^4 - 45x^3 + \frac{45}{2}x^2 - 18x + 9 = 0 ,$$

Its roots are: 1) $x^2 + 1 = 0, x_1 = \sqrt{-1} = i; x_2 = -\sqrt{-1} = -i$; 2) $x^2 + 3 = 0, x_3 = \sqrt{3}i; x_4 = -\sqrt{3}i$;

3) $x^2 + 2 = 0, x_5 = \sqrt{2}i; x_6 = -\sqrt{2}i$; 4) $x^2 + \dfrac{3}{2} = 0, x_7 = \sqrt{\dfrac{3}{2}}i; x_8 = -\sqrt{\dfrac{3}{2}}i$; 5) $-2x + 1 = 0, x_9 = \dfrac{1}{2}$.

❷ : when $a = 4$, it means that:

$$c = -39; d = -\dfrac{39}{4}; e = 107; f = \dfrac{107}{4}; g = -108; h = -27; i = 36 \ .$$

Now, the equation is:

$$4x^9 + x^8 - 39x^7 - \dfrac{39}{4}x^6 + 107x^5 + \dfrac{107}{4}x^4 - 108x^3 - 27x^2 + 36x + 9 = 0 \ ,$$

Its roots are: 1) $x^2 - 2 = 0, x_1 = \sqrt{2}; x_2 = -\sqrt{2}$; 2) $x^2 - 6 = 0, x_3 = \sqrt{6}; x_4 = -\sqrt{6}$;

3) $x^2 - 1 = 0, x_5 = 1; x_6 = -1$; 4) $x^2 - \dfrac{3}{4} = 0, x_7 = \dfrac{\sqrt{3}}{2}; x_8 = -\dfrac{\sqrt{3}}{2}$; 5) $4x + 1 = 0, x_9 = -\dfrac{1}{4}$.

… … ∞ … …

So that when $a \in R; a \neq 0; b = 1; c = -\left(7 + 2a^2\right); d = -\left(\dfrac{7}{a} + 2a\right); e = \dfrac{12}{a} + 14a + \dfrac{3a^3}{4}; f = \dfrac{12}{a^2} + 14 + \dfrac{3a^2}{4}$;

$g = -\left(24 + \dfrac{21a^2}{4}\right); h = -\left(\dfrac{24}{a} + \dfrac{21a}{4}\right); i = 9a; j = 9$, the equation

$ax^9 + bx^8 + cx^7 + dx^6 + ex^5 + fx^4 + gx^3 + hx^2 + ix + j = 0$, which has the roots: $x_1 = \sqrt{\dfrac{a}{2}}; x_2 = -\sqrt{\dfrac{a}{2}}$;

$x_3 = \sqrt{\dfrac{3a}{2}}; x_4 = -\sqrt{\dfrac{3a}{2}}; \ x_5 = \sqrt{\dfrac{3}{a}}; x_6 = -\sqrt{\dfrac{3}{a}}; \ x_7 = \sqrt{\dfrac{4}{a}}; x_8 = -\sqrt{\dfrac{4}{a}}; \ x_9 = -\dfrac{1}{a}$.

I call it one of the Dinbakish Theorems.

①: According to the equation $ax^9 + bx^8 + cx^7 + dx^6 + ex^5 + fx^4 + gx^3 + hx^2 + ix + j = 0$, can you fill in the blanks with the correct numbers for it?

[1]: $(1) + (\ ) + (-18) + (\ ) + (-23) + (\ ) + (\ ) + (\ ) + (\ ) + (-216) = 0$;

[2]: $(\ ) + (-16) + (\ ) + (\ ) + (\ ) + (92) + (\ ) + (\ ) + (\ ) + (-216) = 0$;

[3]: $(\ ) + (\ ) + (\ ) + (144) + (\ ) + (\ ) + (\ ) + (516) + (\ ) + (-216) = 0$;

$[4]$: $(\ )+(81)+(\ )+(\ )+(\ )+(207)+(\ )+(\ )+(\ )+(-216)=0$;

$[5]$: $(\ )+(\ )+(\ )+(486)+(\ )+(\ )+(\ )+(774)+(\ )+(-216)=0$;

$[6]$: $(\ )+(\ )+(-2304)+(\ )+(-736)+(\ )+(\ )+(\ )+(\ )+(-216)=0$;

$[7]$: $(-512)+(\ )+(\ )+(\ )+(\ )+(\ )+(2064)+(\ )+(\ )+(-216)=0$;

$[8]$: $(\ )+(-6561)+(\ )+(\ )+(\ )+(\ )+(1863)+(\ )+(648)+(-216)=0$;

$[9]$: $(\ )+(\ )+(\ )+(13122)+(\ )+(\ )+(\ )+(\ )+(-648)+(-216)=0$;

$a=(\ );b=(\ );c=(\ );d=(\ );e=(\ );f=(\ );g=(\ );h=(\ );i=(\ )$;

$x_1=(\ );x_2=(\ );x_3=(\ );x_4=(\ );x_5=(\ );x_6=(\ );x_7=(\ );x_8=(\ );x_9=(\ )$.

ii. One of the forms of this equation can be:

$$\left(x^2+ax+a\right)\left(x^2-\frac{a}{3}\right)\left(x^2-\frac{2a}{3}\right)\left(x^2-\frac{27}{a^2}\right)(ax-2)=0 ,$$

It expands:

$$ax^9+\left(a^2-2\right)x^8-\left(2a+\frac{27}{a}\right)x^7+\left(\frac{54}{a^2}-27-a^3\right)x^6+\left(\frac{54}{a}+2a^2-\frac{7a^3}{9}\right)x^5+\left(\frac{14a^2}{9}+23a+\frac{2a^4}{9}\right)x^4-\left(54+\frac{4a^3}{9}-21a-\frac{2a^4}{9}\right)x^3-\left(42+\frac{4a^3}{9}+6a^2\right)x^2+\left(12a-6a^2\right)x+12a=0 ,$$

And $b=a^2-2;c=-\left(2a+\dfrac{27}{a}\right);d=\dfrac{54}{a^2}-27-a^3;e=\dfrac{54}{a}+2a^2-\dfrac{7a^3}{9};f=\dfrac{14a^2}{9}+23a+\dfrac{2a^4}{9}$;

$$g=-\left(54+\frac{4a^3}{9}-21a-\frac{2a^4}{9}\right);h=-\left(42+\frac{4a^3}{9}+6a^2\right);i=12a-6a^2;j=12a ;$$

The equation follows the real number a to be changed.

❶ : when $a=3$, it means that:

$b=7;c=-15;d=-48;e=15;f=113;g=15;h=-108;i=-18;j=36$.

Now, the equation is:

$$3x^9 + 7x^8 - 15x^7 - 48x^6 + 15x^5 + 113x^4 + 15x^3 - 108x^2 - 18x + 36 = 0 ,$$

Its roots are: 1) $x^2 + 3x + 3 = 0, x_1 = \dfrac{-3 + \sqrt{3}i}{2}; x_2 = \dfrac{-3 - \sqrt{3}i}{2}$; 2) $x^2 - 1 = 0, x_3 = 1; x_4 = -1$;

3) $x^2 - 2 = 0, x_5 = \sqrt{2}; x_6 = -\sqrt{2}$; 4) $x^2 - 3 = 0, x_7 = \sqrt{3}; x_8 = -\sqrt{3}$; 5) $3x - 2 = 0, x_9 = \dfrac{2}{3}$.

②: when $a = -3$, it means that:

$$b = 7; c = 15; d = 6; e = 21; f = -49; g = -87; h = -84; i = -18; j = -36 .$$

Now, the equation is:

$$-3x^9 + 7x^8 + 15x^7 + 6x^6 + 21x^5 - 49x^4 - 87x^3 - 84x^2 - 18x - 36 = 0 ,$$

Its roots are: 1) $x^2 - 3x - 3 = 0, x_1 = \dfrac{3 + \sqrt{21}}{2}; x_2 = \dfrac{3 - \sqrt{21}}{2}$; 2) $x^2 + 1 = 0, x_3 = i; x_4 = -i$;

3) $x^2 + 2 = 0, x_5 = \sqrt{2}i; x_6 = -\sqrt{2}i$; 4) $x^2 - 3 = 0, x_7 = \sqrt{3}; x_8 = -\sqrt{3}$; 5) $-3x - 2 = 0, x_9 = -\dfrac{2}{3}$.

… … ∞ … …

So that when $a \in R; a \neq 0; b = a^2 - 2; c = -\left(2a + \dfrac{27}{a}\right); d = \dfrac{54}{a^2} - 27 - a^3; e = \dfrac{54}{a} + 2a^2 - \dfrac{7a^3}{9}; f = \dfrac{14a^2}{9} + 23a + \dfrac{2a^4}{9}$;

$g = -\left(54 + \dfrac{4a^3}{9} - 21a - \dfrac{2a^4}{9}\right); h = -\left(42 + \dfrac{4a^3}{9} + 6a^2\right); i = 12a - 6a^2; j = 12a$, the equation

$ax^9 + bx^8 + cx^7 + dx^6 + ex^5 + fx^4 + gx^3 + hx^2 + ix + j = 0$, which has the roots: $x_1 = \dfrac{-a + \sqrt{a^2 - 4a}}{2}$;

$x_2 = \dfrac{-a - \sqrt{a^2 - 4a}}{2}$; $x_3 = \sqrt{\dfrac{a}{3}}; x_4 = -\sqrt{\dfrac{a}{3}}$; $x_5 = \sqrt{\dfrac{2a}{3}}; x_6 = -\sqrt{\dfrac{2a}{3}}$; $x_7 = \sqrt{\dfrac{27}{a^2}}; x_8 = -\sqrt{\dfrac{27}{a^2}}$;

$x_9 = \dfrac{2}{a}$. I call it one of the Dinbakish Theorems.

②: According to the equation $ax^9 + bx^8 + cx^7 + dx^6 + ex^5 + fx^4 + gx^3 + hx^2 + ix + j = 0$, can you fill in the blanks with the correct numbers for it?

$[1]$: $(\ )+(\ )+(-20)+(\ )+(137)+(\ )+(-382)+(\ )+(\ )+(-360)=0$;

$[2]$: $(\ )+(-16)+(\ )+(\ )+(\ )+(\ )+(\ )+(764)+(\ )+(-360)=0$;

$[3]$: $(\ )+(\ )+(\ )+(160)+(\ )+(-548)+(\ )+(\ )+(\ )+(-360)=0$;

$[4]$: $(\ )+(-81)+(\ )+(\ )+(\ )+(\ )+(\ )+(1146)+(\ )+(-360)=0$;

$[5]$: $(\ )+(\ )+(\ )+(540)+(\ )+(-1233)+(\ )+(\ )+(\ )+(-360)=0$;

$[6]$: $(512)+(\ )+(\ )+(\ )+(4384)+(\ )+(\ )+(\ )+(720)+(-360)=0$;

$[7]$: $(\ )+(\ )+(2560)+(\ )+(\ )+(\ )+(3056)+(\ )+(\ )+(-360)=0$;

$[8]$: $(19683)+(\ )+(\ )+(\ )+(33291)+(\ )+(\ )+(\ )+(1080)+(-360)=0$;

$[9]$: $(\ )+(\ )+(43740)+(\ )+(\ )+(\ )+(10314)+(\ )+(\ )+(-360)=0$;

$a=(\ );b=(\ );c=(\ );d=(\ );e=(\ );f=(\ );g=(\ );h=(\ );i=(\ )$;

$x_1=(\ );x_2=(\ );x_3=(\ );x_4=(\ );x_5=(\ );x_6=(\ );x_7=(\ );x_8=(\ );x_9=(\ )$.

(6) : An Equation

$$ax^{10}+bx^9+cx^8+dx^7+ex^6+fx^5+gx^4+hx^3+ix^2+jx+k=0 ,$$

$$a\in R;b\in R;c\in R;d\in R;e\in R;f\in R;g\in R;h\in R;i\in R;j\in R;k\in R;$$

$$a\neq 0;b\neq 0;c\neq 0;d\neq 0;e\neq 0;f\neq 0;g\neq 0;h\neq 0;i\neq 0;j\neq 0;k\neq 0 ;$$

One of the forms of this equation can be:

$$\left(x^2-\frac{a}{5}\right)\left(x^2-\frac{4a}{5}\right)\left(x^2-\frac{1}{2a}\right)\left(x^2-\frac{3}{2a}\right)(ax-5)(x+a)=0 ,$$

It expands:

$$ax^{10}+\left(a^2-5\right)x^9-\left(a^2+5a+2\right)x^8-\left(a^2-5\right)\left(a+\tfrac{2}{a}\right)x^7+\left(5a^2+10+\tfrac{4a^3}{25}+2a+\tfrac{3}{4a}\right)x^6+\left(a^2-5\right)\left(\tfrac{4a^2}{25}+2+\tfrac{3}{4a^2}\right)x^5-\left(\tfrac{4a^3}{5}+10a+\tfrac{15}{4a}+\tfrac{8a^2}{25}+\tfrac{3}{4}\right)x^4-\left(a^2-5\right)\left(\tfrac{8a}{25}+\tfrac{3}{4a}\right)x^3+\left(\tfrac{8a^2}{5}+\tfrac{15}{4}+\tfrac{3a}{25}\right)x^2+\tfrac{3}{25}\left(a^2-5\right)x-\tfrac{3a}{5}=0 \quad,$$

And $b=a^2-5;c=-\left(a^2+5a+2\right);d=-\left(a^2-5\right)\left(a+\dfrac{2}{a}\right);e=5a^2+10+\dfrac{4a^3}{25}+2a+\dfrac{3}{4a};$

$$f=\left(a^2-5\right)\left(\dfrac{4a^2}{25}+2+\dfrac{3}{4a^2}\right);g=-\left(\dfrac{4a^3}{5}+10a+\dfrac{15}{4a}+\dfrac{8a^2}{25}+\dfrac{3}{4}\right);h=-\left(a^2-5\right)\left(\dfrac{8a}{25}+\dfrac{3}{4a}\right);i=\dfrac{8a^2}{5}+\dfrac{15}{4}+\dfrac{3a}{25};$$

$$j=\dfrac{3}{25}\left(a^2-5\right);k=-\dfrac{3a}{5}\ .$$

The equation follows the real number a to be changed.

1 : when $a=5$, t means that:

$$b=20;c=-52;d=-108;e=\dfrac{3303}{20};f=\dfrac{603}{5};g=-\dfrac{319}{2};h=-35;i=\dfrac{883}{20};j=\dfrac{12}{5};k=-3\ .$$

Now, the equation is:

$$5x^{10}+20x^9-52x^8-108x^7+\dfrac{3303}{20}x^6+\dfrac{603}{5}x^5-\dfrac{319}{2}x^4-35x^3+\dfrac{887}{20}x^2+\dfrac{12}{5}x-3=0\ ,$$

Its roots are: 1) $x^2-1=0,x_1=1;x_2=-1$; 2) $x^2-4=0,x_3=2;x_4=-2$; 3) $x^2-\dfrac{3}{10}=0,x_5=\sqrt{\dfrac{3}{10}}$;

$x_6=-\sqrt{\dfrac{3}{10}}$; 4) $x^2-\dfrac{1}{10}=0,x_7=\dfrac{\sqrt{10}}{10};x_8=-\dfrac{\sqrt{10}}{10}$; 5) $5x-5=0,x_9=1$; 6) $x+5=0,x_{10}=-5$.

2 : when $a=-5$, it means that:

$$b=20;c=-2;d=108;e=\dfrac{2097}{20};f=\dfrac{603}{5};g=142;h=35;i=\dfrac{863}{20};j=\dfrac{12}{5};k=3\ .$$

Now, the equation is:

$$-5x^{10}+20x^9-2x^8+108x^7+\dfrac{2097}{20}x^6+\dfrac{603}{5}x^5+142x^4+35x^3+\dfrac{863}{20}x^2+\dfrac{12}{5}x+3=0\ ,$$

Its roots are: 1) $x^2+1=0,x_1=i;x_2=-i$; 2) $x^2+4=0,x_3=2i;x_4=-2i$; 3) $x^2+\dfrac{1}{10}=0,x_5=\dfrac{\sqrt{10}}{10}i$;

$x_6 = -\dfrac{\sqrt{10}}{10}i$; 4) $x^2 + \dfrac{3}{10} = 0, x_7 = \sqrt{\dfrac{3}{10}}i; x_8 = -\sqrt{\dfrac{3}{10}}i$; 5) $-5x - 5 = 0, x_9 = -1$; 6) $x - 5 = 0, x_{10} = 5$.

… … ∞ … …

So that when $a \in R; a \neq 0; b = a^2 - 5; c = -\left(a^2 + 5a + 2\right); d = -\left(a^2 - 5\right)\left(a + \dfrac{2}{a}\right); e = 5a^2 + 10 + \dfrac{4a^3}{25} + 2a + \dfrac{3}{4a};$

$f = \left(a^2 - 5\right)\left(\dfrac{4a^2}{25} + 2 + \dfrac{3}{4a^2}\right); g = -\left(\dfrac{4a^3}{5} + 10a + \dfrac{15}{4a} + \dfrac{8a^2}{25} + \dfrac{3}{4}\right); h = -\left(a^2 - 5\right)\left(\dfrac{8a}{25} + \dfrac{3}{4a}\right); i = \dfrac{8a^2}{5} + \dfrac{15}{4} + \dfrac{3a}{25};$

$j = \dfrac{3}{25}\left(a^2 - 5\right); k = -\dfrac{3a}{5}$, the equation $ax^{10} + bx^9 + cx^8 + dx^7 + ex^6 + fx^5 + gx^4 + hx^3 + ix^2 + jx + k = 0$,

which has the roots: $x_1 = \sqrt{\dfrac{a}{5}}; x_2 = -\sqrt{\dfrac{a}{5}}$; $x_3 = \sqrt{\dfrac{4a}{5}}; x_4 = -\sqrt{\dfrac{4a}{5}}; x_5 = \sqrt{\dfrac{1}{2a}}; x_6 = -\sqrt{\dfrac{1}{2a}}; x_7 = \sqrt{\dfrac{3}{2a}}$;

$x_8 = -\sqrt{\dfrac{3}{2a}}$; $x_9 = \dfrac{5}{a}; x_{10} = -a$. I call it one of the Dinbakish Theorems.

②: According to the equation $ax^{10} + bx^9 + cx^8 + dx^7 + ex^6 + fx^5 + gx^4 + hx^3 + ix^2 + jx + k = 0$, can you fill in the blanks with correct numbers for it?

$[1]$: $(32) + (\ ) + (\ ) + (\ ) + (-680) + (\ ) + (\ ) + (\ ) + (-1152) + (\ ) + (288) = 0$;

$[2]$: $(\ ) + (\ ) + (32) + (\ ) + (\ ) + (\ ) + (1480) + (\ ) + (\ ) + (\ ) + (288) = 0$;

$[3]$: $(243) + (\ ) + (\ ) + (\ ) + (-2295) + (\ ) + (\ ) + (\ ) + (-1728) + (\ ) + (288) = 0$;

$[4]$: $(\ ) + (\ ) + (162) + (\ ) + (\ ) + (\ ) + (3330) + (\ ) + (\ ) + (\ ) + (288) = 0$;

$[5]$: $(\ ) + (-7) + (\ ) + (70) + (\ ) + (\ ) + (\ ) + (\ ) + (\ ) + (-168) + (288) = 0$;

$[6]$: $(\ ) + (7) + (\ ) + (\ ) + (\ ) + (245) + (\ ) + (-350) + (\ ) + (\ ) + (288) = 0$;

$[7]$: $(\ ) + (\ ) + (\ ) + (\ ) + (\ ) + (-7840) + (\ ) + (\ ) + (\ ) + (-336) + (288) = 0$;

$[8]$: $(\ ) + (\ ) + (\ ) + (-8920) + (\ ) + (\ ) + (\ ) + (-2800) + (\ ) + (\ ) + (288) = 0$;

$[9]$: $(\ ) + (\ ) + (13122) + (\ ) + (\ ) + (-59535) + (\ ) + (\ ) + (-5184) + (\ ) + (288) = 0$;

$[10]$: $(\ ) + (-1835008) + (\ ) + (\ ) + (-348160) + (\ ) + (94720) + (\ ) + (\ ) + (\ ) + (288) = 0$;

111

$a=(\ );b=(\ );c=(\ );d=(\ );e=(\ );f=(\ );g=(\ );h=(\ );i=(\ );j=(\ )\ ;$

$x_1=(\ );x_2=(\ );x_3=(\ );x_4=(\ );x_5=(\ );x_6=(\ );x_7=(\ );x_8=(\ );x_9=(\ );x_{10}=(\ )\ .$

ii: One of the forms of this equation can be:

$$\left(ax^2+ax+1\right)\left(x^2-\frac{1}{2a}\right)\left(x^2-\frac{3}{2a}\right)\left(x^2-2a\right)\left(x^2-4a^2\right)=0\ ,$$

It expands:

$$ax^{10}+ax^9-\left(4a^3+2a^2+1\right)x^8-\left(4a^3+2a^2+2\right)x^7+\left(8a^4+4a^2+2a-\frac{5}{4a}\right)x^6+\left(8a^4+8a^2+4a+\frac{3}{4a}\right)x^5-\left(8a^3-5a-\frac{5}{2}-\frac{3}{4a^2}\right)x^4-\left(16a^3+3a+\frac{3}{2}\right)x^3-\left(10a^2+3+\frac{3}{2a}\right)x^2+6a^2x+6a=0\ ,$$

And $b=a;c=-\left(4a^3+2a^2+1\right);d=-\left(4a^3+2a^2+2\right);e=8a^4+4a^2+2a-\dfrac{5}{4a};f=8a^4+8a^2+4a+\dfrac{3}{4a}$;

$g=-\left(8a^3-5a-\dfrac{5}{2}-\dfrac{3}{4a^2}\right);h=-\left(16a^3+3a+\dfrac{3}{2}\right);i=-\left(10a^2+3+\dfrac{3}{2a}\right);j=6a^2;k=6a$.

The equation follows the real number a to be changed.

1: when $a=\dfrac{1}{2}$, it means that:

$b=\dfrac{1}{2};c=-2;d=-3;e=0;f=6;g=7;h=-5;i=-\dfrac{17}{2};j=\dfrac{3}{2};k=3$.

Now, the equation is:

$$\frac{1}{2}x^{10}+\frac{1}{2}x^9-2x^8-3x^7+6x^5+7x^4-5x^3-\frac{17}{2}x^2+\frac{3}{2}x+3=0\ ,$$

Its roots are: 1) $\dfrac{1}{2}x^2+\dfrac{1}{2}x+1=0,x_1=\dfrac{-1+\sqrt{7}i}{2};x_2=\dfrac{-1-\sqrt{7}i}{2}$; 2) $x^2-1=0,x_3=1;x_4=-1$;

3) $x^2-3=0,x_5=\sqrt{3};x_6=-\sqrt{3}$; 4) $x^2-1=0,x_7=1;x_8=-1$; 5) $x^2-1=0,x_9=1;x_{10}=-1$.

2: when $a=-\dfrac{1}{2}$, it means that:

$b=-\dfrac{1}{2};c=-1;d=-2;e=3;f=-1;g=4;h=2;i=-\dfrac{5}{2};j=\dfrac{3}{2};k=-3$.

112

Now, the equation is:

$$-\frac{1}{2}x^{10} - \frac{1}{2}x^9 - x^8 - 2x^7 + 3x^6 - x^5 + 4x^4 + 2x^3 - \frac{5}{2}x^2 + \frac{3}{2}x - 3 = 0 \ ,$$

Its roots are: 1) $-\frac{1}{2}x^2 - \frac{1}{2}x + 1 = 0, x_1 = 1; x_2 = -2$; 2) $x^2 + 1 = 0, x_3 = i; x_4 = -i$; 3) $x^2 + 3 = 0$,

$x_5 = \sqrt{3}i; x_6 = -\sqrt{3}i$; 4) $x^2 + 1 = 0, x_7 = i; x_8 = -i$; 5) $x^2 - 1 = 0, x_9 = 1; x_{10} = -1$.

... ∞ ...

So that when $a \in R; a \neq 0; b = a; c = -(4a^3 + 2a^2 + 1); d = -(4a^3 + 2a^2 + 2); e = 8a^4 + 4a^2 + 2a - \frac{5}{4a}; f = 8a^4 + 8a^2 + 4a + \frac{3}{4a}$;

$g = -\left(8a^3 - 5a - \frac{5}{2} - \frac{3}{4a^2}\right); h = -\left(16a^3 + 3a + \frac{3}{2}\right); i = -\left(10a^2 + 3 + \frac{3}{2a}\right); j = 6a^2; k = 6a$;

the equation $ax^{10} + bx^9 + cx^8 + dx^7 + ex^6 + fx^5 + gx^4 + hx^3 + ix^2 + jx + k = 0$, which has the roots:

$$x_1 = -\frac{1}{2} + \sqrt{\frac{1}{4} - \frac{1}{a}}; x_2 = -\frac{1}{2} - \sqrt{\frac{1}{4} - \frac{1}{a}}; \ x_3 = \sqrt{\frac{1}{2a}}; x_4 = -\sqrt{\frac{1}{2a}}; \ x_5 = \sqrt{\frac{3}{2a}}; x_6 = -\sqrt{\frac{3}{2a}};$$

$x_7 = \sqrt{2a}; x_8 = -\sqrt{2a}; \ x_9 = 2a; x_{10} = -2a$. I call it one of the Dinbakish Theorems.

②: According to the equation $ax^{10} + bx^9 + cx^8 + dx^7 + ex^6 + fx^5 + gx^4 + hx^3 + ix^2 + jx + k = 0$, can you fill in the blanks with the correct numbers for it?

[1] : ( ) + (−5) + ( ) + ( ) + ( ) + (105) + ( ) + ( ) + ( ) + (−600) + (720) = 0 ;

[2] : ( ) + ( ) + ( ) + (30) + ( ) + ( ) + ( ) + (−530) + ( ) + ( ) + (720) = 0 ;

[3] : (3125) + ( ) + ( ) + ( ) + (1875) + ( ) + ( ) + ( ) + (−2580) + ( ) + (720) = 0 ;

[4] : ( ) + ( ) + (7500) + ( ) + ( ) + ( ) + (−5800) + ( ) + ( ) + ( ) + (720) = 0 ;

[5] : ( ) + ( ) + (15552) + ( ) + ( ) + ( ) + (−8352) + ( ) + ( ) + ( ) + (720) = 0 ;

[6] : (7776) + ( ) + ( ) + ( ) + (3240) + ( ) + ( ) + ( ) + (−3096) + ( ) + (720) = 0 ;

$[7]:(\ )+(-2560)+(\ )+(\ )+(\ )+(3360)+(\ )+(\ )+(\ )+(-1200)+(720)=0$;

$[8]:(1024)+(\ )+(\ )+(3840)+(\ )+(\ )+(\ )+(-4240)+(\ )+(\ )+(720)=0$;

$[9]:(\ )+(-2560)+(\ )+(\ )+(\ )+(\ )+(-3712)+(4240)+(\ )+(\ )+(720)=0$;

$[10]:(\ )+(-98415)+(\ )+(-65610)+(\ )+(\ )+(-18792)+(\ )+(\ )+(-1800)+(720)=0$;

$a=(\ );b=(\ );c=(\ );d=(\ );e=(\ );f=(\ );g=(\ );h=(\ );i=(\ );j=(\ )$;

$x_1=(\ );x_2=(\ );x_3=(\ );x_4=(\ );x_5=(\ );x_6=(\ );x_7=(\ );x_8=(\ );x_9=(\ );x_{10}=(\ )$.

About the other equations $ax^{11}+bx^{10}+cx^9+dx^8+ex^7+fx^6+gx^5+hx^4+ix^3+jx^2+kx+l=0$,
$ax^{12}+bx^{11}+cx^{10}+dx^9+ex^8+fx^7+gx^6+hx^5+ix^4+jx^3+kx^2+lx+m=0$,
$ax^{13}+bx^{12}+cx^{11}+dx^{10}+ex^9+fx^8+gx^7+hx^6+ix^5+jx^4+kx^3+lx^2+mx+n=0$, …, I have no time to show them in the Volume 2. But they have the general forms like these:

(1) : $a_{2n+1}x^{2n+1}+a_{2n}x^{2n}+a_{2n-1}x^{2n-1}+...+a_2x^2+a_1x+a_0=0$,

One of the forms of this equation can be:

$(x^2-k_1)(x^2-k_2)(x^2-k_3)...(x^2-k_n)(a_{2n+1}x-k_1)=0$.

(2) : $a_{2n+2}x^{2n+2}+a_{2n+1}x^{2n+1}+a_{2n}x^{2n}+...+a_2x^2+a_1x+a_0=0$,

One of the forms of this equation can be:

$(x^2-k_1)(x^2-k_2)(x^2-k_3)...(x^2-k_n)(a_{2n}x-k_1)(x+k_2)=0$.

($n=1,2,3,...,\infty;k_1\in R;k_2\in R;k_3\in R;...;k_n\in R;k_1\neq0;k_2\neq0;k_3\neq0;...;k_n\neq0$)

2: How to apply the Dinbakish Theorems for making something great?

(1) : An Equation

$$ax^5 + bx^4 + cx^3 + dx^2 + ex + f = 0 ,$$

$$a \in R; b \in R; c \in R; d \in R; e \in R; f \in R; a \neq 0; b \neq 0; c \neq 0; d \neq 0; e \neq 0; f \neq 0;$$

One of the forms of this equation can be:

$$\left(x^2 - 1\right)\left(x^2 - 9\right)\left(ax - 1\right) = 0 ,$$

It expands:

$$ax^5 - x^4 - 10ax^3 + 10x^2 + 9ax - 9 = 0 ,$$

The equation follows the real number a to be changed.

❶ : when $a = 1$, the equation is:

$$x^5 - x^4 - 10x^3 + 10x^2 + 9x - 9 = 0 ,$$

Its roots are:

$$x_1 = 1; x_2 = -1; x_3 = 3; x_4 = -3; x_5 = 1;$$

And the equation becomes a matrix:

$[1]$: $(1) + (-1) + (-10) + (10) + (9) + (-9) = 0$;

$[2]$: $(-1) + (-1) + (10) + (10) + (-9) + (-9) = 0$;

$[3]$: $(1) + (-1) + (-10) + (10) + (9) + (-9) = 0$;

$[4]$: $(243) + (-81) + (-270) + (90) + (27) + (-9) = 0$;

$[5]$: $(-243) + (-81) + (270) + (90) + (-27) + (-9) = 0$.

❷ : when $a = 2$, the equation is:

$$2x^5 - x^4 - 20x^3 + 10x^2 + 18x - 9 = 0 ,$$

Its roots are:

$$x_1 = 1; x_2 = -1; x_3 = 3; x_4 = -3; x_5 = \frac{1}{2};$$

And the equation becomes a matrix:

$[1]$: $(2)+(-1)+(-20)+(10)+(18)+(-9)=0$;

$[2]$: $(-2)+(-1)+(20)+(10)+(-18)+(-9)=0$;

$[3]$: $(486)+(-81)+(-540)+(90)+(54)+(-9)=0$;

$[4]$: $(-486)+(-81)+(540)+(90)+(-54)+(-9)=0$;

$[5]$: $\left(\dfrac{1}{16}\right)+\left(-\dfrac{1}{16}\right)+\left(-\dfrac{5}{2}\right)+\left(\dfrac{5}{2}\right)+(9)+(-9)=0$.

❸ : when $a=3$, the equation is:

$$3x^5 - x^4 - 30x^3 + 10x^2 + 27x - 9 = 0 ,$$

Its roots are:

$$x_1 = 1; x_2 = -1; x_3 = 3; x_4 = -3; x_5 = \dfrac{1}{3};$$

And the equation becomes a matrix:

$[1]$: $(3)+(-1)+(-30)+(10)+(27)+(-9)=0$;

$[2]$: $(-3)+(-1)+(30)+(10)+(-27)+(-9)=0$;

$[3]$: $(729)+(-81)+(-810)+(90)+(81)+(-9)=0$;

$[4]$: $(-729)+(-81)+(810)+(90)+(-81)+(-9)=0$;

$[5]$: $\left(\dfrac{1}{81}\right)+\left(-\dfrac{1}{81}\right)+\left(-\dfrac{10}{9}\right)+\left(\dfrac{10}{9}\right)+(9)+(-9)=0$.

…… ∞ ……

(2) : An Equation

$$ax^6 + bx^5 + cx^4 + dx^3 + ex^2 + fx + g = 0$$

$a \in R; b \in R; c \in R; d \in R; e \in R; f \in R; g \in R; a \neq 0; b \neq 0; c \neq 0; d \neq 0; e \neq 0; f \neq 0; g \neq 0;$

One of the forms of this equation can be:

$$\left(x^2 - 4\right)\left(x^2 - 9\right)\left(x + 5\right)\left(ax - 1\right) = 0 \ ,$$

It expands:

$$ax^6 + \left(5a-1\right)x^5 - \left(5+13a\right)x^4 - 13\left(5a-1\right)x^3 + \left(65+36a\right)x^2 + 36\left(5a-1\right)x - 180 = 0 \ ,$$

The equation follows the real number a to be changed.

❶ : when $a = 1$, the equation is:

$$x^6 + 4x^5 - 18x^4 - 52x^3 + 101x^2 + 144x - 180 = 0 \ ,$$

Its roots are:

$$x_1 = 2; x_2 = -2; x_3 = 3; x_4 = -3; x_5 = -5; x_6 = 1 \ ;$$

And the equation becomes a matrix:

$[1]$: $(1) + (4) + (-18) + (-52) + (101) + (144) + (-180) = 0$;

$[2]$: $(15625) + (-12500) + (-11250) + (6500) + (2525) + (-720) + (-180) = 0$;

$[3]$: $(64) + (128) + (-288) + (-416) + (404) + (288) + (-180) = 0$;

$[4]$: $(64) + (-128) + (-288) + (416) + (404) + (-288) + (-180) = 0$;

$[5]$: $(729) + (972) + (-1458) + (-1404) + (909) + (432) + (-180) = 0$;

$[6]$: $(729) + (-972) + (-1458) + (1404) + (909) + (-432) + (-180) = 0$.

❷ : when $a = 2$, the equation is:

$$2x^6 + 9x^5 - 31x^4 - 117x^3 + 137x^2 + 324x - 180 = 0 \ ,$$

Its roots are:

$$x_1 = 2; x_2 = -2; x_3 = 3; x_4 = -3; x_5 = -5; x_6 = \frac{1}{2} \ ;$$

And the equation becomes a matrix:

$[1]:\left(\dfrac{1}{32}\right)+\left(\dfrac{9}{32}\right)+\left(-\dfrac{31}{16}\right)+\left(-\dfrac{117}{8}\right)+\left(\dfrac{137}{4}\right)+(162)+(-180)=0$;

$[2]:(31250)+(-28125)+(-19375)+(14625)+(3425)+(-1620)+(-180)=0$;

$[3]:(128)+(288)+(-496)+(-936)+(548)+(648)+(-180)=0$;

$[4]:(128)+(-288)+(-496)+(936)+(548)+(-648)+(-180)=0$;

$[5]:(1458)+(2187)+(-2511)+(-3159)+(1233)+(972)+(-180)=0$;

$[6]:(1458)+(-2187)+(-2511)+(3159)+(1233)+(-972)+(-180)=0$.

❸ : when $a=3$, the equation is:

$$3x^6+14x^5-44x^4-182x^3+173x^2+504x-180=0 ,$$

Its roots are:

$$x_1=2; x_2=-2; x_3=3; x_4=-3; x_5=-5; x_6=\dfrac{1}{3} ;$$

And the equation becomes a matrix:

$[1]:\left(\dfrac{1}{243}\right)+\left(\dfrac{14}{243}\right)+\left(-\dfrac{44}{81}\right)+\left(-\dfrac{182}{27}\right)+\left(\dfrac{173}{9}\right)+(168)+(-180)=0$;

$[2]:(46875)+(-43750)+(-27500)+(22750)+(4325)+(-2520)+(-180)=0$;

$[3]:(192)+(448)+(-704)+(-1456)+(692)+(1008)+(-180)=0$;

$[4]:(192)+(-448)+(-704)+(1456)+(692)+(-1008)+(-180)=0$;

$[5]:(2187)+(3402)+(-3564)+(-4914)+(1557)+(1512)+(-180)=0$;

$[6]:(2187)+(-3402)+(-3564)+(4914)+(1557)+(-1512)+(-180)=0$.

(3) : An Equation

$$ax^7 + bx^6 + cx^5 + dx^4 + ex^3 + fx^2 + gx + h = 0 \ ,$$

$$a \in R; b \in R; c \in R; d \in R; e \in R; f \in R; g \in R; h \in R;$$

$$a \neq 0; b \neq 0; c \neq 0; d \neq 0; e \neq 0; f \neq 0; g \neq 0; h \neq 0 \ .$$

One of the forms of this equation can be:

$$\left(x^2 - 1\right)\left(x^2 - 4\right)\left(x^2 - 9\right)\left(ax - 1\right) = 0 \ ,$$

It expands:

$$ax^7 - x^6 - 14ax^5 + 14x^4 + 49ax^3 - 49x^2 - 36ax + 36 = 0 \ .$$

The equation follows the real number a to be changed.

❶ : when $a = 1$, the equation is:

$$x^7 - x^6 - 14x^5 + 14x^4 + 49x^3 - 49x^2 - 36x + 36 = 0 \ ,$$

Its roots are:

$$x_1 = 1; x_2 = -1; x_3 = 2; x_4 = -2; x_5 = 3; x_6 = -3; x_7 = 1 \ ;$$

And the equation becomes a matrix:

$[1]$: $(1) + (-1) + (-14) + (14) + (49) + (-49) + (-36) + (36) = 0$;

$[2]$: $(-1) + (-1) + (14) + (14) + (-49) + (-49) + (36) + (36) = 0$;

$[3]$: $(1) + (-1) + (-14) + (14) + (49) + (-49) + (-36) + (36) = 0$;

$[4]$: $(128) + (-64) + (-448) + (224) + (392) + (-196) + (-72) + (36) = 0$;

$[5]$: $(-128) + (-64) + (448) + (224) + (-392) + (-196) + (72) + (36) = 0$;

$[6]: (2187) + (-729) + (-3402) + (1134) + (1323) + (-441) + (-108) + (36) = 0$;

$[7]: (-2187) + (-729) + (3402) + (1134) + (-1323) + (-441) + (108) + (36) = 0$.

❷ : when $a = 2$, the equation is:

$$2x^7 - x^6 - 28x^5 + 14x^4 + 98x^3 - 49x^2 - 72x + 36 = 0 \ ,$$

Its roots are:

$$x_1 = 1; x_2 = -1; x_3 = 2; x_4 = -2; x_5 = 3; x_6 = -3; x_7 = \frac{1}{2} \ ;$$

And the equation becomes a matrix:

$[1]: (2) + (-1) + (-28) + (14) + (98) + (-49) + (-72) + (36) = 0$;

$[2]: (-2) + (-1) + (28) + (14) + (-98) + (-49) + (72) + (36) = 0$;

$[3]: (256) + (-64) + (-896) + (224) + (784) + (-196) + (-144) + (36) = 0$;

$[4]: (-256) + (-64) + (896) + (224) + (-784) + (-196) + (144) + (36) = 0$;

$[5]: (4374) + (-729) + (-6804) + (1134) + (2646) + (-441) + (-216) + (36) = 0$;

$[6]: (-4374) + (-729) + (6804) + (1134) + (-2646) + (-441) + (216) + (36) = 0$;

$[7]: \left(\frac{1}{64}\right) + \left(-\frac{1}{64}\right) + \left(-\frac{7}{8}\right) + \left(\frac{7}{8}\right) + \left(\frac{49}{4}\right) + \left(-\frac{49}{4}\right) + (-36) + (36) = 0$.

❸ : when $a = 3$, the equation is:

$$3x^7 - x^6 - 42x^5 + 14x^4 + 147x^3 - 49x^2 - 108x + 36 = 0 \ ,$$

Its roots are:

$$x_1 = 1; x_2 = -1; x_3 = 2; x_4 = -2; x_5 = 3; x_6 = -3; x_7 = \frac{1}{3} \ ;$$

And the equation becomes a matrix:

$[1] : (3)+(-1)+(-42)+(14)+(147)+(-49)+(-108)+(36)=0$;

$[2] : (-3)+(-1)+(42)+(14)+(-147)+(-49)+(108)+(36)=0$;

$[3] : (384)+(-64)+(-1344)+(224)+(1176)+(-196)+(-216)+(36)=0$;

$[4] : (-384)+(-64)+(1344)+(224)+(-1176)+(-196)+(216)+(36)=0$;

$[5] : (6561)+(-729)+(-10206)+(1134)+(3969)+(-441)+(-324)+(36)=0$;

$[6] : (-6561)+(-729)+(10206)+(1134)+(-3969)+(-441)+(324)+(36)=0$;

$[7] : \left(\dfrac{1}{729}\right)+\left(-\dfrac{1}{729}\right)+\left(-\dfrac{14}{81}\right)+\left(\dfrac{14}{81}\right)+\left(\dfrac{49}{9}\right)+\left(-\dfrac{49}{9}\right)+(-36)+(36)=0$.

… … ∞ … …

(4) : An Equation

$$ax^8 + bx^7 + cx^6 + dx^5 + ex^4 + fx^3 + gx^2 + hx + i = 0 \ ,$$

$$a \in R; b \in R; c \in R; d \in R; e \in R; f \in R; g \in R; h \in R; i \in R;$$

$$a \neq 0; b \neq 0; c \neq 0; d \neq 0; e \neq 0; f \neq 0; g \neq 0; h \neq 0; i \neq 0 \ .$$

One of the forms of this equation can be:

$$\left(x^2 -1\right)\left(x^2 -4\right)\left(x^2 - 9\right)(x - 5)(ax-1) = 0 \ ,$$

It expands:

$$ax^8 -(5a+1)x^7 -(14a-5)x^6 +14(5a+1)x^5 +(49a-70)x^4 -4a(5a+1)x^3 +(245-36a)x^2 +36(5a+1)x -180 = 0 \ .$$

The equation follows the real number a to be changed.

❶ : when $a = 1$, the equation is:

$$x^8 - 6x^7 - 9x^6 + 84x^5 - 21x^4 - 294x^3 + 209x^2 + 216x - 180 = 0 ,$$

Its roots are:

$$x_1 = 1; x_2 = -1; x_3 = 2; x_4 = -2; x_5 = 3; x_6 = -3; x_7 = 5; x_8 = 1 ;$$

And the equation becomes a matrix:

$[1]$: $(1) + (-6) + (-9) + (84) + (-21) + (-294) + (209) + (216) + (-180) = 0$;

$[2]$: $(1) + (6) + (-9) + (-84) + (-21) + (294) + (209) + (-216) + (-180) = 0$;

$[3]$: $(256) + (-768) + (-576) + (2688) + (-336) + (-2352) + (836) + (432) + (-180) = 0$;

$[4]$: $(256) + (768) + (-576) + (-2688) + (-336) + (2352) + (836) + (-432) + (-180) = 0$;

$[5]$: $(6561) + (-13122) + (-6561) + (20412) + (-1701) + (-7938) + (1881) + (648) + (-180) = 0$;

$[6]$: $(6561) + (13122) + (-6561) + (-20412) + (-1701) + (7938) + (1881) + (-648) + (-180) = 0$;

$[7]$: $(1) + (-6) + (-9) + (84) + (-21) + (-294) + (209) + (216) + (-180) = 0$;

$[8]$: $(390625) + (-468750) + (-140625) + (262500) + (-13125) + (-36750) + (5225) + (1080) + (-180) = 0$.

❷ : when $a = 2$, the equation is:

$$2x^8 - 11x^7 - 23x^6 + 154x^5 + 28x^4 - 539x^3 + 173x^2 + 396x - 180 = 0 ,$$

Its roots are:

$$x_1 = 1; x_2 = -1; x_3 = 2; x_4 = -2; x_5 = 3; x_6 = -3; x_7 = 5; x_8 = \frac{1}{2} ;$$

And the equation becomes a matrix:

$[1]$: $(2) + (-11) + (-23) + (154) + (28) + (-539) + (173) + (396) + (-180) = 0$;

$[2]$: $(2) + (11) + (-23) + (-154) + (28) + (539) + (173) + (-396) + (-180) = 0$;

$[3]: (512)+(-1408)+(-1472)+(4928)+(448)+(-4312)+(692)+(792)+(-180)=0$;

$[4]: (512)+(1408)+(-1472)+(-4928)+(448)+(4312)+(692)+(-792)+(-180)=0$;

$[5]: (13122)+(-24057)+(-16767)+(37422)+(2268)+(-14553)+(1557)+(1188)+(-180)=0$;

$[6]: (13122)+(24057)+(-16767)+(-37422)+(2268)+(14553)+(1557)+(-1188)+(-180)=0$;

$[7]: \left(\dfrac{1}{128}\right)+\left(-\dfrac{11}{128}\right)+\left(-\dfrac{23}{64}\right)+\left(\dfrac{77}{16}\right)+\left(\dfrac{7}{4}\right)+\left(-\dfrac{539}{8}\right)+\left(\dfrac{173}{4}\right)+(198)+(-180)=0$;

$[8]: (781250)+(-859375)+(-359375)+(481250)+(17500)+(-67375)+(4325)+(1980)+(-180)=0$.

❸: when $a=3$, the equation is:

$3x^8-16x^7-37x^6+224x^5+77x^4-784x^3+137x^2+576x-180=0$,

Its roots are:

$x_1=1; x_2=-1; x_3=2; x_4=-2; x_5=3; x_6=-3; x_7=5; x_8=\dfrac{1}{3}$;

And the equation becomes a matrix:

$[1]: (3)+(-16)+(-37)+(224)+(77)+(-784)+(137)+(576)+(-180)=0$;

$[2]: (3)+(16)+(-37)+(-224)+(77)+(784)+(137)+(-576)+(-180)=0$;

$[3]: (768)+(-2048)+(-2368)+(7168)+(1232)+(-6272)+(548) | (1152)+(-180)-0$;

$[4]: (768)+(2048)+(-2368)+(-7168)+(1232)+(6272)+(548)+(-1152)+(-180)=0$;

$[5]: (19683)+(-34992)+(-26973)+(54432)+(6237)+(-21168)+(1233)+(1728)+(-180)=0$;

$[6]: (19683)+(34992)+(-26973)+(-54432)+(6237)+(21168)+(1233)+(-1728)+(-180)=0$;

$[7]: \left(\dfrac{1}{2187}\right)+\left(-\dfrac{16}{2187}\right)+\left(-\dfrac{37}{729}\right)+\left(\dfrac{224}{243}\right)+\left(\dfrac{77}{81}\right)+\left(-\dfrac{784}{27}\right)+\left(\dfrac{137}{9}\right)+(192)+(-180)=0$;

$[8]$: $(1171875)+(-1250000)+(-578125)+(700000)+(48125)+(-98000)+(3425)+(2880)+(-180)=0$.

… … ∞ … …

(5) : An Equation

$$ax^9 + bx^8 + cx^7 + dx^6 + ex^5 + fx^4 + gx^3 + hx^2 + ix + j = 0 \ ,$$

$$a \in R; b \in R; c \in R; d \in R; e \in R; f \in R; g \in R; h \in R; i \in R; j \in R;$$

$$a \neq 0; b \neq 0; c \neq 0; d \neq 0; e \neq 0; f \neq 0; g \neq 0; h \neq 0; i \neq 0; j \neq 0 \ .$$

One of the forms of this equation can be:

$$\left(x^2-1\right)\left(x^2-4\right)\left(x^2-9\right)\left(x^2-16\right)\left(ax-1\right)=0 \ ,$$

It expands:

$$ax^9 - x^8 - 30ax^7 + 30x^6 + 273ax^5 - 273x^4 - 820ax^3 + 820x^2 + 576ax - 576 = 0 \ .$$

The equation follows the real number a to be changed.

1 : when $a = 1$, the equation is:

$$x^9 - x^8 - 30x^7 + 30x^6 + 273x^5 - 273x^4 - 820x^3 + 820x^2 + 576x - 576 = 0 \ ,$$

Its roots are:

$$x_1 = 1; x_2 = -1; x_3 = 2; x_4 = -2; x_5 = 3; x_6 = -3; x_7 = 4; x_8 = -4; x_9 = 1 \ ;$$

And the equation becomes a matrix:

$[1]$: $(1)+(-1)+(-30)+(30)+(273)+(-273)+(-820)+(820)+(576)+(-576)=0$;

$[2]$: $(-1)+(-1)+(30)+(30)+(-273)+(-273)+(820)+(820)+(-576)+(-576)=0$;

$[3]$: $(1)+(-1)+(-30)+(30)+(273)+(-273)+(-820)+(820)+(576)+(-576)=0$;

$[4]$: $(512)+(-256)+(-3840)+(1920)+(8736)+(-4368)+(-6560)+(3280)+(1152)+(-576)=0$;

$[5]$: $(-512)+(-256)+(3840)+(1920)+(-8736)+(-4368)+(6560)+(3280)+(-1152)+(-576)=0$;

$[6]$: $(19683)+(-6561)+(-65610)+(21870)+(66339)+(-22113)+(-22140)+(7380)+(1728)+(-576)=0$;

$[7]$: $(-19683)+(-6561)+(65610)+(21870)+(-66339)+(-22113)+(22140)+(7380)+(-1728)+(-576)=0$;

$[8]$: $(262144)+(-65536)+(-491520)+(122880)+(279552)+(-69888)+(-52480)+(13120)+(2304)+(-576)=0$;

$[9]$: $(-262144)+(-65536)+(491520)+(122880)+(-279552)+(-69888)+(52480)+(13120)+(-2304)+(-576)=0$.

❷ : when $a=2$, the equation is:

$2x^9 - x^8 - 60x^7 + 30x^6 + 546x^5 - 273x^4 - 1640x^3 + 820x^2 + 1152x - 576 = 0$,

Its roots are:

$x_1 = 1; x_2 = -1; x_3 = 2; x_4 = -2; x_5 = 3; x_6 = -3; x_7 = 4; x_8 = -4; x_9 = \dfrac{1}{2};$

And the equation becomes a matrix:

$[1]$: $(2)+(-1)+(-60)+(30)+(546)+(-273)+(-1640)+(820)+(1152)+(-576)=0$;

$[2]$: $(-2)+(-1)+(60)+(30)+(-564)+(-273)+(1640)+(820)+(-1152)+(-576)=0$;

$[3]$: $(1024)+(-256)+(-7680)+(1920)+(17472)+(-4368)+(-13120)+(3280)+(2304)+(-576)=0$;

$[4]$: $(-1024)+(-256)+(7680)+(1920)+(-17472)+(-4368)+(13120)+(3280)+(-2304)+(-576)=0$;

$[5]$: $(39366)+(-6561)+(-131220)+(21870)+(132678)+(-22113)+(-44280)+(7380)+(3456)+(-576)=0$;

$[6]$: $(-39366)+(-6561)+(131220)+(21870)+(-132678)+(-22113)+(44280)+(7380)+(-3456)+(-576)=0$;

$[7]$: $(524288)+(-65536)+(-983040)+(122880)+(559104)+(-69888)+(-104960)+(13120)+(4608)+(-576)=0$;

$[8]$: $(-524288)+(-65536)+(983040)+(122880)+(-559104)+(-69888)+(104960)+(13120)+(-4608)+(-576)=0$;

$[9]$: $\left(\dfrac{1}{256}\right)+\left(-\dfrac{1}{256}\right)+\left(-\dfrac{15}{32}\right)+\left(\dfrac{15}{32}\right)+\left(\dfrac{273}{16}\right)+\left(-\dfrac{273}{16}\right)+(-205)+(205)+(576)+(-576)=0$.

❸ : when $a = 3$, the equation is:

$$3x^9 - x^8 - 90x^7 + 30x^6 + 819x^5 - 273x^4 - 2460x^3 + 820x^2 + 1728x - 576 = 0 \ ,$$

Its roots are:

$$x_1 = 1; x_2 = -1; x_3 = 2; x_4 = -2; x_5 = 3; x_6 = -3; x_7 = 4; x_8 = -4; x_9 = \frac{1}{3};$$

And the equation becomes a matrix:

$[1]$: $(3) + (-1) + (-90) + (30) + (819) + (-273) + (-2460) + (820) + (1728) + (-576) = 0$;

$[2]$: $(-3) + (-1) + (90) + (30) + (-819) + (-273) + (2460) + (820) + (-1728) + (-576) = 0$;

$[3]$: $(1536) + (-256) + (-11520) + (1920) + (26208) + (-4368) + (-19680) + (3280) + (3456) + (-576) = 0$;

$[4]$: $(-1536) + (-256) + (11520) + (1920) + (-26208) + (-4368) + (19680) + (3280) + (-3456) + (-576) = 0$;

$[5]$: $(59049) + (-6561) + (-196830) + (21870) + (199017) + (-22113) + (-66420) + (7380) + (5184) + (-576) = 0$;

$[6]$: $(-59049) + (-6561) + (196830) + (21870) + (-199017) + (-22113) + (66420) + (7380) + (-5184) + (-576) = 0$;

$[7]$: $(786432) + (-65536) + (-1474560) + (122880) + (838656) + (-69888) + (-157440) + (13120) + (6912) + (-576) = 0$;

$[8]$: $(-786432) + (-65536) + (1474560) + (122880) + (-838656) + (-69888) + (157440) + (13120) + (-6912) + (-576) = 0$;

$[9]$: $\left(\frac{1}{6561}\right) + \left(-\frac{1}{6561}\right) + \left(-\frac{10}{243}\right) + \left(\frac{10}{243}\right) + \left(\frac{91}{27}\right) + \left(-\frac{91}{27}\right) + \left(-\frac{820}{9}\right) + \left(\frac{820}{9}\right) + (576) + (-576) = 0$.

… … ∞ … …

(6) : An Equation

$$ax^{10} + bx^9 + cx^8 + dx^7 + ex^6 + fx^5 + gx^4 + hx^3 + ix^2 + jx + k = 0 \ ,$$

$$a \in R; b \in R; c \in R; d \in R; e \in R; f \in R; g \in R; h \in R; i \in R; j \in R; k \in R;$$

$$a \neq 0; b \neq 0; c \neq 0; d \neq 0; e \neq 0; f \neq 0; g \neq 0; h \neq 0; i \neq 0; j \neq 0; k \neq 0;$$

One of the forms of this equation can be:

$$\left(x^2 - 1\right)\left(x^2 - 4\right)\left(x^2 - 9\right)\left(x^2 - 16\right)\left(x - 5\right)\left(ax - 1\right) = 0 \ ,$$

It expands:

$$ax^{10} - (5a+1)x^9 - (30a-5)x^8 + (150a+30)x^7 + (273a-150)x^6 - (1365a+273)x^5 - (820a-1365)x^4 + (4100a+820)x^3 + (576a-4100)x^2 - (2880a+576)x + 2880 = 0 \ .$$

The equation follows the real number a to be changed.

1 : when $a = 1$,the equation is:

$$x^{10} - 6x^9 - 25x^8 + 180x^7 + 123x^6 - 1638x^5 + 545x^4 + 4920x^3 - 3524x^2 + 3456x + 2880 = 0 \ ,$$

Its roots are:

$$x_1 = 1; x_2 = -1; x_3 = 2; x_4 = -2; x_5 = 3; x_6 = -3; x_7 = 4; x_8 = -4; x_9 = 5; x_{10} = 1 \ ;$$

And the equation becomes a matrix:

$[1]$: $(1)+(-6)+(-25)+(180)+(123)+(-1638)+(545)+(4920)+(-3524)+(3456)+(2880) = 0$;

$[2]$: $(1)+(6)+(-25)+(-180)+(123)+(1638)+(545)+(-4920)+(-3524)+(3456)+(2880) = 0$;

$[3]$: $(1)+(-6)+(-25)+(180)+(123)+(-1638)+(545)+(4920)+(-3524)+(3456)+(2880) = 0$;

$[4]$: $(1024)+(-3072)+(-6400)+(23040)+(7872)+(-52416)+(8720)+(39360)+(-14096)+(-6912)+(2880) = 0$;

$[5]$: $(1024)+(3072)+(-6400)+(-23040)+(7872)+(52416)+(8720)+(39360)+(-14096)+(6912)+(2880) = 0$;

$[6]$: $(59049)+(-118098)+(-164025)+(393660)+(89667)+(-398034)+(44145)+(132840)+(-31716)+(-10368)+(2880) = 0$;

$[7]$: $(59049)+(118098)+(-164025)+(-393660)+(89667)+(398034)+(44145)+(-132840)+(-31716)+(10368)+(2880) = 0$;

$[8]$: $(1048576)+(-1572864)+(-1638400)+(2949120)+(503808)+(-1677312)+(139520)+(314880)+(-56384)+(-13824)+(2880) = 0$;

$[9]$: $(1048576)+(1572864)+(-1638400)+(-2949120)+(503808)+(1677312)+(139520)+(-314880)+(-56384)+(13824)+(2880) = 0$;

$[10]$: $(9765625)+(-11718750)+(-9765625)+(14062500)+(1921875)+(-5118750)+(340625)+(615000)+(-88100)+(-17280)+(2880)=0$.

2 : when $a=2$, the equation is:

$$2x^{10}-11x^9-55x^8+330x^7+396x^6-3003x^5-275x^4+9020x^3-2948x^2-6336x+2880=0 \ ,$$

Its roots are:

$$x_1=1; x_2=-1; x_3=2; x_4=-2; x_5=3; x_6=-3; x_7=4; x_8=-4; x_9=5; x_{10}=\frac{1}{2};$$

And the equation becomes a matrix:

$[1]$: $(2)+(-11)+(-55)+(330)+(396)+(-3003)+(-275)+(9020)+(-2948)+(-6336)+(2880)=0$;

$[2]$: $(2)+(11)+(-55)+(-330)+(396)+(3003)+(-275)+(-9020)+(-2948)+(6336)+(2880)=0$;

$[3]$: $(2048)+(-5632)+(-14080)+(42240)+(25344)+(-96096)+(-4400)+(72160)+(-11792)+(-12672)+(2880)=0$;

$[4]$: $(2048)+(5632)+(-14080)+(-42240)+(25344)+(96096)+(-4400)+(-72160)+(-11792)+(12672)+(2880)=0$;

$[5]$: $(118098)+(-216513)+(-360855)+(721710)+(288684)+(-729729)+(-22275)+(243540)+(-26532)+(-19008)+(2880)=0$;

$[6]$: $(118098)+(216513)+(-360855)+(-721710)+(288684)+(729729)+(-22275)+(-243540)+(-26532)+(19008)+(2880)=0$;

$[7]$: $(2097152)+(-2883584)+(-3604480)+(5406720)+(1622016)+(-3075072)+(-70400)+(577280)+(-47168)+(-25344)+(2880)=0$;

$[8]$: $(2097152)+(2883584)+(-3604480)+(-5406720)+(16622016)+(3075072)+(-70400)+(-577280)+(-47168)+(25344)+(2880)=0$;

$[9]$: $(19531250)+(-21484375)+(-21484375)+(25781250)+(6187500)+(-9384375)+(-171875)+(1127500)+(-73700)+(-31680)+(2880)=0$;

$[10]$: $\left(\frac{1}{512}\right)+\left(-\frac{11}{512}\right)+\left(-\frac{55}{256}\right)+\left(\frac{330}{128}\right)+\left(\frac{396}{64}\right)+\left(-\frac{3003}{32}\right)+\left(-\frac{275}{16}\right)+\left(\frac{2255}{2}\right)+(-737)+(-3168)+(2880)=0$.

3 : when $a=3$, the equation is:

$$3x^{10}-16x^9-85x^8+480x^7+669x^6-4368x^5-1095x^4+13120x^3-2372x^2-9216x+2880=0 \ ,$$

Its roots are:

128

$$x_1 = 1; x_2 = -1; x_3 = 2; x_4 = -2; x_5 = 3; x_6 = -3; x_7 = 4; x_8 = -4; x_9 = 5; x_{10} = \frac{1}{3};$$

And the equation becomes a matrix:

$[1]$: $(3) + (-16) + (-85) + (480) + (669) + (-4368) + (-1095) + (13120) + (-2372) + (-9216) + (2880) = 0$;

$[2]$: $(3) + (16) + (-85) + (-480) + (669) + (4368) + (-1095) + (-13120) + (-2372) + (9216) + (2880) = 0$;

$[3]$: $(3072) + (-8192) + (-21760) + (61440) + (42816) + (-139776) + (-17520) + (104960) + (-9488) + (-18432) + (2880) = 0$;

$[4]$: $(3072) + (8192) + (-21760) + (-61440) + (42816) + (139776) + (-17520) + (-104960) + (-9488) + (18432) + (2880) = 0$;

$[5]$: $(177147) + (-314928) + (-557685) + (1049760) + (487701) + (-1061424) + (-88695) + (354240) + (-21348) + (-27648) + (2880) = 0$;

$[6]$: $(177147) + (314928) + (-557685) + (-1049760) + (487701) + (1061424) + (-88695) + (-354240) + (-21348) + (27648) + (2880) = 0$;

$[7]$: $(3145728) + (-4194304) + (-5570560) + (7864320) + (2740224) + (-4472832) + (-280320) + (839680) + (-37952) + (-36864) + (2880) = 0$;

$[8]$: $(3145728) + (4194304) + (-5570560) + (-7864320) + (2740224) + (4472832) + (-280320) + (-839680) + (-37952) + (36864) + (2880) = 0$;

$[9]$: $(29296875) + (-31250000) + (-33203125) + (37500000) + (10453125) + (-13650000) + (-684375) + (1640000) + (-59300) + (-46080) + (2880) = 0$;

$[10]$: $\left(\frac{1}{19683}\right) + \left(-\frac{16}{19683}\right) + \left(-\frac{85}{6561}\right) + \left(\frac{160}{729}\right) + \left(\frac{223}{243}\right) + \left(-\frac{1456}{81}\right) + \left(-\frac{365}{27}\right) + \left(\frac{13120}{27}\right) + \left(-\frac{2372}{9}\right) + (-3072) + (2880) = 0$.

…… ∞ ……

3: The Multivariate Funn Equations

i: $a_n \left(\dfrac{x}{y}\right)^n + a_{n-1} \left(\dfrac{x}{y}\right)^{n-1} + a_{n-2} \left(\dfrac{x}{y}\right)^{n-2} + ... + a_2 \left(\dfrac{x}{y}\right)^2 + a_1 \left(\dfrac{x}{y}\right) + a_0 = 0$;

ii: $a_n \left(\dfrac{x}{yz}\right)^n + a_{n-1} \left(\dfrac{x}{yz}\right)^{n-1} + a_{n-2} \left(\dfrac{x}{yz}\right)^{n-2} + ... + a_2 \left(\dfrac{x}{yz}\right)^2 + a_1 \left(\dfrac{x}{yz}\right) + a_0 = 0$;

iii: $a_n \left(\dfrac{xz}{y}\right)^n + a_{n-1} \left(\dfrac{xz}{y}\right)^{n-1} + a_{n-2} \left(\dfrac{xz}{y}\right)^{n-2} + ... + a_2 \left(\dfrac{xz}{y}\right)^2 + a_1 \left(\dfrac{xz}{y}\right) + a_0 = 0$;

iv: $a_n\left(\dfrac{xz}{yv}\right)^n + a_{n-1}\left(\dfrac{xz}{yv}\right)^{n-1} + a_{n-2}\left(\dfrac{xz}{yv}\right)^{n-2} + \ldots + a_2\left(\dfrac{xz}{yv}\right)^2 + a_1\left(\dfrac{xz}{yv}\right) + a_0 = 0$;

$$y = a_1 x + a_0 ; z = a_2 x + a_0 ; v = a_1 x + a_2 ; n = 2,3,4,\ldots,\infty;$$

$\boxed{(1)}$: A Multivariate Funn Equation

$$a\left(\dfrac{x}{y}\right)^2 + b\left(\dfrac{x}{y}\right) + c = 0 \ , \ y = ax + b \ ;$$

$$a \in R; b \in R; c \in R; a \neq 0; b \neq 0; c \neq 0 \ .$$

This Multivariate Funn Equation can be:

$$\left(\dfrac{x}{y}\right)^2 + \dfrac{b}{a}\left(\dfrac{x}{y}\right) + \dfrac{c}{a} = 0 \ ; \ \left(\dfrac{x}{y} + \dfrac{b}{2a}\right)^2 = \dfrac{b^2 - 4ac}{4a^2} \ ;$$

In which it has two parts:

Part 1: $\dfrac{x}{y} = \dfrac{-b + \sqrt{b^2 - 4ac}}{2a}$; Part 2: $\dfrac{x}{y} = \dfrac{-b - \sqrt{b^2 - 4ac}}{2a}$.

In the part 1:

$$x = \left(\dfrac{-b + \sqrt{b^2 - 4ac}}{2a}\right) y \ ; \ y = ax + b \ ;$$

$$y = a\left(\dfrac{-b + \sqrt{b^2 - 4ac}}{2a}\right) y + b \ ; \ \left(\dfrac{2 + b - \sqrt{b^2 - 4ac}}{2}\right) y = b \ ;$$

$$y_1 = \dfrac{2b}{2 + b - \sqrt{b^2 - 4ac}} \ ; \text{ and } x_1 = \dfrac{-b^2 + b\sqrt{b^2 - 4ac}}{2a + ab - a\sqrt{b^2 - 4ac}} \ .$$

In the part 2:

130

$$x = \left(\frac{-b - \sqrt{b^2 - 4ac}}{2a} \right) y \; ; \; y = ax + b \; ;$$

$$y = a\left(\frac{-b - \sqrt{b^2 - 4ac}}{2a} \right) y + b \; ; \; \left(\frac{2 + b + \sqrt{b^2 - 4ac}}{2} \right) y = b \; ;$$

$$y_2 = \frac{2b}{2 + b + \sqrt{b^2 - 4ac}} \; ; \text{ and } \; x_2 = \frac{-b^2 - b\sqrt{b^2 - 4ac}}{2a + ab + a\sqrt{b^2 - 4ac}} \; .$$

$$(2) : \text{ A Multivariate Funn Equation}$$

$$a\left(\frac{x}{yz} \right)^2 + b\left(\frac{x}{yz} \right) + c = 0 \; ; \; y = bx + c; z = ax + c;$$

$$a \in R; b \in R; c \in R; a \neq 0; b \neq 0; c \neq 0 \; .$$

This Multivariate Funn Equation can be:

$$\left(\frac{x}{yz} \right)^2 + \frac{b}{a}\left(\frac{x}{yz} \right) + \frac{c}{a} = 0 \; ; \; \left(\frac{x}{yz} + \frac{b}{2a} \right)^2 = \frac{b^2 - 4ac}{4a^2} \; .$$

In which it has two parts:

Part 1: $\dfrac{x}{yz} = \dfrac{-b + \sqrt{b^2 - 4ac}}{2a}$; Part 2: $\dfrac{x}{yz} = \dfrac{-b - \sqrt{b^2 - 4ac}}{2a}$.

In the part 1:

$$\frac{x}{(bx + c)(ax + c)} = \frac{-b + \sqrt{b^2 - 4ac}}{2a} \; ; \; \frac{x}{abx^2 + (bc + ac)x + c^2} = \frac{-b + \sqrt{b^2 - 4ac}}{2a} \; ;$$

$$\frac{ab\left(-b + \sqrt{b^2 - 4ac}\right)}{2a} x^2 + \frac{(bc + ac)\left(-b + \sqrt{b^2 - 4ac}\right)}{2a} x + \frac{c^2\left(-b + \sqrt{b^2 - 4ac}\right)}{2a} = x \; ;$$

$$\frac{ab\left(-b+\sqrt{b^2-4ac}\right)}{2a}x^2+\frac{(bc+ac)\left(-b+\sqrt{b^2-4ac}\right)-2a}{2a}x+\frac{c^2\left(-b+\sqrt{b^2-4ac}\right)}{2a}=0\ ;$$

$$x^2+\frac{(bc+ac)\left(-b+\sqrt{b^2-4ac}\right)-2a}{ab\left(-b+\sqrt{b^2-4ac}\right)}x+\frac{c^2\left(-b+\sqrt{b^2-4ac}\right)}{ab\left(-b+\sqrt{b^2-4ac}\right)}=0\ ;$$

$$\left(x+\frac{(bc+ac)\left(-b+\sqrt{b^2-4ac}\right)-2a}{2ab\left(-b+\sqrt{b^2-4ac}\right)}\right)^2-\frac{\left[(bc+ac)\left(-b+\sqrt{b^2-4ac}\right)-2a\right]^2}{\left[2ab\left(-b+\sqrt{b^2-4ac}\right)\right]^2}+\frac{c^2\left(-b+\sqrt{b^2-4ac}\right)}{ab\left(-b+\sqrt{b^2-4ac}\right)}=0\ ;$$

$$\left(x+\frac{(bc+ac)\left(-b+\sqrt{b^2-4ac}\right)-2a}{2ab\left(-b+\sqrt{b^2-4ac}\right)}\right)^2=\frac{\left[(bc+ac)\left(-b+\sqrt{b^2-4ac}\right)-2a\right]^2}{\left[2ab\left(-b+\sqrt{b^2-4ac}\right)\right]^2}-\frac{c^2\left(-b+\sqrt{b^2-4ac}\right)4ab\left(-b+\sqrt{b^2-4ac}\right)}{\left[2ab\left(-b+\sqrt{b^2-4ac}\right)\right]^2}\ ;$$

$$\left(x+\frac{(bc+ac)\left(-b+\sqrt{b^2-4ac}\right)-2a}{2ab\left(-b+\sqrt{b^2-4ac}\right)}\right)^2=\frac{\left[(bc+ac)\left(-b+\sqrt{b^2-4ac}\right)-2a\right]^2-c^2\left(-b+\sqrt{b^2-4ac}\right)4ab\left(-b+\sqrt{b^2-4ac}\right)}{\left[2ab\left(-b+\sqrt{b^2-4ac}\right)\right]^2}\ ;$$

$$x_1=\frac{-\left[(bc+ac)\left(-b+\sqrt{b^2-4ac}\right)-2a\right]+\sqrt{\left[(bc+ac)\left(-b+\sqrt{b^2-4ac}\right)-2a\right]^2-c^2\left(-b+\sqrt{b^2-4ac}\right)4ab\left(-b+\sqrt{b^2-4ac}\right)}}{2ab\left(-b+\sqrt{b^2-4ac}\right)}\ ;$$

$$x_2=\frac{-\left[(bc+ac)\left(-b+\sqrt{b^2-4ac}\right)-2a\right]-\sqrt{\left[(bc+ac)\left(-b+\sqrt{b^2-4ac}\right)-2a\right]^2-c^2\left(-b+\sqrt{b^2-4ac}\right)4ab\left(-b+\sqrt{b^2-4ac}\right)}}{2ab\left(-b+\sqrt{b^2-4ac}\right)}\ ;$$

$$y_1=bx_1+c;\ y_2=bx_2+c;$$

$$y_1=b\left\{\frac{-\left[(bc+ac)\left(-b+\sqrt{b^2-4ac}\right)-2a\right]+\sqrt{\left[(bc+ac)\left(-b+\sqrt{b^2-4ac}\right)-2a\right]^2-c^2\left(-b+\sqrt{b^2-4ac}\right)4ab\left(-b+\sqrt{b^2-4ac}\right)}}{2ab\left(-b+\sqrt{b^2-4ac}\right)}\right\}+c\ ;$$

$$y_2=b\left\{\frac{-\left[(bc+ac)\left(-b+\sqrt{b^2-4ac}\right)-2a\right]-\sqrt{\left[(bc+ac)\left(-b+\sqrt{b^2-4ac}\right)-2a\right]^2-c^2\left(-b+\sqrt{b^2-4ac}\right)4ab\left(-b+\sqrt{b^2-4ac}\right)}}{2ab\left(-b+\sqrt{b^2-4ac}\right)}\right\}+c\ ;$$

$$z_1 = ax_1 + c; z_2 = ax_2 + c;$$

$$z_1 = a\left\{\frac{-\left[(bc+ac)\left(-b+\sqrt{b^2-4ac}\right)-2a\right]+\sqrt{\left[(bc+ac)\left(-b+\sqrt{b^2-4ac}\right)-2a\right]^2-c^2\left(-b+\sqrt{b^2-4ac}\right)4ab\left(-b+\sqrt{b^2-4ac}\right)}}{2ab\left(-b+\sqrt{b^2-4ac}\right)}\right\}+c \ ;$$

$$z_2 = a\left\{\frac{-\left[(bc+ac)\left(-b+\sqrt{b^2-4ac}\right)-2a\right]+\sqrt{\left[(bc+ac)\left(-b+\sqrt{b^2-4ac}\right)-2a\right]^2-c^2\left(-b+\sqrt{b^2-4ac}\right)4ab\left(-b+\sqrt{b^2-4ac}\right)}}{2ab\left(-b+\sqrt{b^2-4ac}\right)}\right\}+c \ .$$

In the part 2:

$$\frac{x}{(bx+c)(ax+c)}=\frac{-b-\sqrt{b^2-4ac}}{2a} \ ; \quad \frac{x}{abx^2+(bc+ac)x+c^2}=\frac{-b-\sqrt{b^2-4ac}}{2a} \ ;$$

$$\frac{ab\left(-b-\sqrt{b^2-4ac}\right)}{2a}x^2+\frac{(bc+ac)\left(-b-\sqrt{b^2-4ac}\right)}{2a}x+\frac{c^2\left(-b-\sqrt{b^2-4ac}\right)}{2a}=x \ ;$$

$$\frac{ab\left(-b-\sqrt{b^2-4ac}\right)}{2a}x^2+\frac{(bc+ac)\left(-b-\sqrt{b^2-4ac}\right)-2a}{2a}x+\frac{c^2\left(-b-\sqrt{b^2-4ac}\right)}{2a}=0 \ ;$$

$$x^2+\frac{(bc+ac)\left(-b-\sqrt{b^2-4ac}\right)-2a}{ab\left(-b-\sqrt{b^2-4ac}\right)}x+\frac{c^2\left(-b-\sqrt{b^2-4ac}\right)}{ab\left(-b-\sqrt{b^2-4ac}\right)}=0 \ ;$$

$$\left(x+\frac{(bc+ac)\left(-b-\sqrt{b^2-4ac}\right)-2a}{2ab\left(-b-\sqrt{b^2-4ac}\right)}\right)^2-\frac{\left[(bc+ac)\left(-b-\sqrt{b^2-4ac}\right)-2a\right]^2}{\left[2ab\left(-b-\sqrt{b^2-4ac}\right)\right]^2}+\frac{c^2\left(-b-\sqrt{b^2-4ac}\right)}{ab\left(-b-\sqrt{b^2-4ac}\right)}=0 \ ;$$

$$\left(x+\frac{(bc+ac)\left(-b-\sqrt{b^2-4ac}\right)-2a}{2ab\left(-b-\sqrt{b^2-4ac}\right)}\right)^2=\frac{\left[(bc+ac)\left(-b-\sqrt{b^2-4ac}\right)-2a\right]^2}{\left[2ab\left(-b-\sqrt{b^2-4ac}\right)\right]^2}-\frac{c^2\left(-b-\sqrt{b^2-4ac}\right)4ab\left(-b-\sqrt{b^2-4ac}\right)}{\left[2ab\left(-b-\sqrt{b^2-4ac}\right)\right]^2} \ ;$$

$$\left(x + \frac{(bc+ac)\left(-b-\sqrt{b^2-4ac}\right)-2a}{2ab\left(-b-\sqrt{b^2-4ac}\right)} \right)^2 = \frac{\left[(bc+ac)\left(-b-\sqrt{b^2-4ac}\right)-2a\right]^2 - c^2\left(-b-\sqrt{b^2-4ac}\right)4ab\left(-b-\sqrt{b^2-4ac}\right)}{\left[2ab\left(-b-\sqrt{b^2-4ac}\right)\right]^2} \ ;$$

$$x_3 = \frac{-\left[(bc+ac)\left(-b-\sqrt{b^2-4ac}\right)-2a\right] + \sqrt{\left[(bc+ac)\left(-b-\sqrt{b^2-4ac}\right)-2a\right]^2 - c^2\left(-b-\sqrt{b^2-4ac}\right)4ab\left(-b-\sqrt{b^2-4ac}\right)}}{2ab\left(-b-\sqrt{b^2-4ac}\right)} \ ;$$

$$x_4 = \frac{-\left[(bc+ac)\left(-b-\sqrt{b^2-4ac}\right)-2a\right] - \sqrt{\left[(bc+ac)\left(-b-\sqrt{b^2-4ac}\right)-2a\right]^2 - c^2\left(-b-\sqrt{b^2-4ac}\right)4ab\left(-b-\sqrt{b^2-4ac}\right)}}{2ab\left(-b-\sqrt{b^2-4ac}\right)} \ ;$$

$$y_3 = bx_3 + c; y_4 = bx_4 + c;$$

$$y_3 = b\left\{ \frac{-\left[(bc+ac)\left(-b-\sqrt{b^2-4ac}\right)-2a\right] + \sqrt{\left[(bc+ac)\left(-b-\sqrt{b^2-4ac}\right)-2a\right]^2 - c^2\left(-b-\sqrt{b^2-4ac}\right)4ab\left(-b-\sqrt{b^2-4ac}\right)}}{2ab\left(-b-\sqrt{b^2-4ac}\right)} \right\} + c \ ;$$

$$y_4 = b\left\{ \frac{-\left[(bc+ac)\left(-b-\sqrt{b^2-4ac}\right)-2a\right] - \sqrt{\left[(bc+ac)\left(-b-\sqrt{b^2-4ac}\right)-2a\right]^2 - c^2\left(-b-\sqrt{b^2-4ac}\right)4ab\left(-b-\sqrt{b^2-4ac}\right)}}{2ab\left(-b-\sqrt{b^2-4ac}\right)} \right\} + c \ ;$$

$$z_3 = ax_3 + c; z_4 = ax_4 + c;$$

$$z_3 = a\left\{ \frac{-\left[(bc+ac)\left(-b-\sqrt{b^2-4ac}\right)-2a\right] + \sqrt{\left[(bc+ac)\left(-b-\sqrt{b^2-4ac}\right)-2a\right]^2 - c^2\left(-b-\sqrt{b^2-4ac}\right)4ab\left(-b-\sqrt{b^2-4ac}\right)}}{2ab\left(-b-\sqrt{b^2-4ac}\right)} \right\} + c \ ;$$

$$z_4 = a\left\{ \frac{-\left[(bc+ac)\left(-b-\sqrt{b^2-4ac}\right)-2a\right] - \sqrt{\left[(bc+ac)\left(-b-\sqrt{b^2-4ac}\right)-2a\right]^2 - c^2\left(-b-\sqrt{b^2-4ac}\right)4ab\left(-b-\sqrt{b^2-4ac}\right)}}{2ab\left(-b-\sqrt{b^2-4ac}\right)} \right\} + c \ .$$

$$(3) : \text{A Multivariate Funn Equation}$$

$$a\left(\frac{xy}{z}\right)^2 + b\left(\frac{xy}{z}\right) + c = 0 \; ; \; y = bx + c; z = ax + c;$$

$$a \in R; b \in R; c \in R; a \neq 0; b \neq 0; c \neq 0;$$

This Multivariate Funn Equation can be:

$$\left(\frac{xy}{z}\right)^2 + \frac{b}{a}\left(\frac{xy}{z}\right) + \frac{c}{a} = 0 \; ; \; \left(\frac{xy}{z} + \frac{b}{2a}\right)^2 = \frac{b^2 - 4ac}{4a^2} \; ;$$

In which it has two parts:

Part 1: $\dfrac{xy}{z} = \dfrac{-b + \sqrt{b^2 - 4ac}}{2a}$; Part 2: $\dfrac{xy}{z} = \dfrac{-b - \sqrt{b^2 - 4ac}}{2a}$.

In the part 1:

$$\frac{x(bx + c)}{ax + c} = \frac{-b + \sqrt{b^2 - 4ac}}{2a} \; ; \; bx^2 + cx = a\left(\frac{-b + \sqrt{b^2 - 4ac}}{2a}\right)x + c\left(\frac{-b + \sqrt{b^2 - 4ac}}{2a}\right) \; ;$$

$$bx^2 - \frac{a\left(-b + \sqrt{b^2 - 4ac}\right) - 2ac}{2a}x - \frac{c\left(-b + \sqrt{b^2 - 4ac}\right)}{2a} = 0 \; ;$$

$$\left(x - \frac{a\left(-b + \sqrt{b^2 - 4ac}\right) - 2ac}{4ab}\right)^2 = \frac{\left[a\left(-b + \sqrt{b^2 - 4ac}\right) - 2ac\right]^2}{16a^2b^2} + \frac{c\left(-b + \sqrt{b^2 - 4ac}\right)}{2ab} \; ;$$

$$\left(x - \frac{a\left(-b + \sqrt{b^2 - 4ac}\right) - 2ac}{4ab}\right)^2 = \frac{\left[a\left(-b + \sqrt{b^2 - 4ac}\right) - 2ac\right]^2 + c\left(-b + \sqrt{b^2 - 4ac}\right)8ab}{16a^2b^2} \; ;$$

$$x_1 = \frac{a\left(-b + \sqrt{b^2 - 4ac}\right) - 2ac + \sqrt{\left[a\left(-b + \sqrt{b^2 - 4ac}\right) - 2ac\right]^2 + c\left(-b + \sqrt{b^2 - 4ac}\right)8ab}}{4ab} \; ;$$

$$x_2 = \frac{a\left(-b+\sqrt{b^2-4ac}\right)-2ac-\sqrt{\left[a\left(-b+\sqrt{b^2-4ac}\right)-2ac\right]^2+c\left(-b+\sqrt{b^2-4ac}\right)8ab}}{4ab} \;;$$

$$y_1 = bx_1 + c; y_2 = bx_2 + c;$$

$$y_1 = b\left\{\frac{a\left(-b+\sqrt{b^2-4ac}\right)-2ac+\sqrt{\left[a\left(-b+\sqrt{b^2-4ac}\right)-2ac\right]^2+c\left(-b+\sqrt{b^2-4ac}\right)8ab}}{4ab}\right\}+c \;;$$

$$y_2 = b\left\{\frac{a\left(-b+\sqrt{b^2-4ac}\right)-2ac-\sqrt{\left[a\left(-b+\sqrt{b^2-4ac}\right)-2ac\right]^2+c\left(-b+\sqrt{b^2-4ac}\right)8ab}}{4ab}\right\}+c \;;$$

$$z_1 = ax_1 + c; z_2 = ax_2 + c;$$

$$z_1 = a\left\{\frac{a\left(-b+\sqrt{b^2-4ac}\right)-2ac+\sqrt{\left[a\left(-b+\sqrt{b^2-4ac}\right)-2ac\right]^2+c\left(-b+\sqrt{b^2-4ac}\right)8ab}}{4ab}\right\}+c \;;$$

$$z_2 = a\left\{\frac{a\left(-b+\sqrt{b^2-4ac}\right)-2ac-\sqrt{\left[a\left(-b+\sqrt{b^2-4ac}\right)-2ac\right]^2+c\left(-b+\sqrt{b^2-4ac}\right)8ab}}{4ab}\right\}+c \;.$$

In the part 2:

$$\frac{x(bx+c)}{ax+c} = \frac{-b-\sqrt{b^2-4ac}}{2a} \;;\; bx^2+cx = a\left(\frac{-b-\sqrt{b^2-4ac}}{2a}\right)x+c\left(\frac{-b-\sqrt{b^2-4ac}}{2a}\right) \;;$$

$$bx^2 - \frac{a\left(-b-\sqrt{b^2-4ac}\right)-2ac}{2a}x - \frac{c\left(-b-\sqrt{b^2-4ac}\right)}{2a} = 0 \;;$$

$$\left(x-\frac{a\left(-b-\sqrt{b^2-4ac}\right)-2ac}{4ab}\right)^2=\frac{\left[a\left(-b-\sqrt{b^2-4ac}\right)-2ac\right]^2}{16a^2b^2}+\frac{c\left(-b-\sqrt{b^2-4ac}\right)}{2ab}\ ;$$

$$\left(x-\frac{a\left(-b-\sqrt{b^2-4ac}\right)-2ac}{4ab}\right)^2=\frac{\left[a\left(-b-\sqrt{b^2-4ac}\right)-2ac\right]^2+c\left(-b-\sqrt{b^2-4ac}\right)8ab}{16a^2b^2}\ ;$$

$$x_3=\frac{a\left(-b-\sqrt{b^2-4ac}\right)-2ac+\sqrt{\left[a\left(-b-\sqrt{b^2-4ac}\right)-2ac\right]^2+c\left(-b-\sqrt{b^2-4ac}\right)8ab}}{4ab}\ ;$$

$$x_4=\frac{a\left(-b-\sqrt{b^2-4ac}\right)-2ac-\sqrt{\left[a\left(-b-\sqrt{b^2-4ac}\right)-2ac\right]^2+c\left(-b-\sqrt{b^2-4ac}\right)8ab}}{4ab}\ ;$$

$$y_3=bx_3+c;\ y_4=bx_4+c;$$

$$y_3=b\left\{\frac{a\left(-b-\sqrt{b^2-4ac}\right)-2ac+\sqrt{\left[a\left(-b-\sqrt{b^2-4ac}\right)-2ac\right]^2+c\left(-b-\sqrt{b^2-4ac}\right)8ab}}{4ab}\right\}+c\ ;$$

$$y_4=b\left\{\frac{a\left(-b-\sqrt{b^2-4ac}\right)-2ac-\sqrt{\left[a\left(-b-\sqrt{b^2-4ac}\right)-2ac\right]^2+c\left(-b-\sqrt{b^2-4ac}\right)8ab}}{4ab}\right\}+c\ ;$$

$$z_3=ax_3+c;\ z_4=ax_4+c;$$

$$z_3=a\left\{\frac{a\left(-b-\sqrt{b^2-4ac}\right)-2ac+\sqrt{\left[a\left(-b-\sqrt{b^2-4ac}\right)-2ac\right]^2+c\left(-b-\sqrt{b^2-4ac}\right)8ab}}{4ab}\right\}+c\ ;$$

$$z_4 = a\left\{\frac{a\left(-b-\sqrt{b^2-4ac}\right)-2ac-\sqrt{\left[a\left(-b-\sqrt{b^2-4ac}\right)-2ac\right]^2+c\left(-b-\sqrt{b^2-4ac}\right)8ab}}{4ab}\right\}+c \ .$$

4: The Composite-Multivariate Dinbakish-Funn Equations

i: $a_{2n+2}\left(\dfrac{x}{y}\right)^{(n)j+\frac{i}{k}}+a_{2n+1}\left(\dfrac{x}{y}\right)^{(n)j}+a_{2n}\left(\dfrac{x}{y}\right)^{(n-1)j+\frac{i}{k}}+a_{2n-1}\left(\dfrac{x}{y}\right)^{(n-1)j}+...+a_3\left(\dfrac{x}{y}\right)^{j}+a_2\left(\dfrac{x}{y}\right)^{\frac{i}{k}}+a_1=0$;

ii: $a_{2n+2}\left(\dfrac{x}{y}\right)^{(n)j-\frac{i}{k}}+a_{2n+1}\left(\dfrac{x}{y}\right)^{(n)j}+a_{2n}\left(\dfrac{x}{y}\right)^{(n-1)j-\frac{i}{k}}+a_{2n-1}\left(\dfrac{x}{y}\right)^{(n-1)j}+...+a_3\left(\dfrac{x}{y}\right)^{j}+a_2\left(\dfrac{x}{y}\right)^{-\frac{i}{k}}+a_1=0$;

iii: $a_{2n+2}\left(\dfrac{x}{y}\right)^{(n)(-j)+\frac{i}{k}}+a_{2n+1}\left(\dfrac{x}{y}\right)^{(n)(-j)}+a_{2n}\left(\dfrac{x}{y}\right)^{(n-1)(-j)+\frac{i}{k}}+a_{2n-1}\left(\dfrac{x}{y}\right)^{(n-1)(-j)}+...+a_3\left(\dfrac{x}{y}\right)^{-j}+a_2\left(\dfrac{x}{y}\right)^{\frac{i}{k}}+a_1=0$;

iv: $a_{2n+2}\left(\dfrac{x}{y}\right)^{(n)(-j)-\frac{i}{k}}+a_{2n+1}\left(\dfrac{x}{y}\right)^{(n)(-j)}+a_{2n}\left(\dfrac{x}{y}\right)^{(n-1)(-j)-\frac{i}{k}}+a_{2n-1}\left(\dfrac{x}{y}\right)^{(n-1)(-j)}+...+a_3\left(\dfrac{x}{y}\right)^{-j}+a_2\left(\dfrac{x}{y}\right)^{-\frac{i}{k}}+a_1=0$;

v: $a_{2n+2}\left(\dfrac{x}{y}\right)^{(n)j+\sqrt{i}}+a_{2n+1}\left(\dfrac{x}{y}\right)^{(n)j}+a_{2n}\left(\dfrac{x}{y}\right)^{(n-1)j+\sqrt{i}}+a_{2n-1}\left(\dfrac{x}{y}\right)^{(n-1)j}+...+a_3\left(\dfrac{x}{y}\right)^{j}+a_2\left(\dfrac{x}{y}\right)^{\sqrt{i}}+a_1=0$;

vi: $a_{2n+2}\left(\dfrac{x}{y}\right)^{(n)j-\sqrt{i}}+a_{2n+1}\left(\dfrac{x}{y}\right)^{(n)j}+a_{2n}\left(\dfrac{x}{y}\right)^{(n-1)j-\sqrt{i}}+a_{2n-1}\left(\dfrac{x}{y}\right)^{(n-1)j}+...+a_3\left(\dfrac{x}{y}\right)^{j}+a_2\left(\dfrac{x}{y}\right)^{-\sqrt{i}}+a_1=0$;

vii: $a_{2n+2}\left(\dfrac{x}{y}\right)^{(n)(-j)+\sqrt{i}}+a_{2n+1}\left(\dfrac{x}{y}\right)^{(n)(-j)}+a_{2n}\left(\dfrac{x}{y}\right)^{(n-1)(-j)+\sqrt{i}}+a_{2n-1}\left(\dfrac{x}{y}\right)^{(n-1)(-j)}+...+a_3\left(\dfrac{x}{y}\right)^{-j}+a_2\left(\dfrac{x}{y}\right)^{\sqrt{i}}+a_1=0$;

viii: $a_{2n+2}\left(\dfrac{x}{y}\right)^{(n)(-j)-\sqrt{i}}+a_{2n+1}\left(\dfrac{x}{y}\right)^{(n)(-j)}+a_{2n}\left(\dfrac{x}{y}\right)^{(n-1)(-j)-\sqrt{i}}+a_{2n-1}\left(\dfrac{x}{y}\right)^{(n-1)(-j)}+...+a_3\left(\dfrac{x}{y}\right)^{-j}+a_2\left(\dfrac{x}{y}\right)^{-\sqrt{i}}+a_1=0$;

ix: $a_{2n+2}\left(\dfrac{x}{y}\right)^{(n)j+\frac{\sqrt{i}}{k}}+a_{2n+1}\left(\dfrac{x}{y}\right)^{(n)j}+a_{2n}\left(\dfrac{x}{y}\right)^{(n-1)j+\frac{\sqrt{i}}{k}}+a_{2n-1}\left(\dfrac{x}{y}\right)^{(n-1)j}+...+a_3\left(\dfrac{x}{y}\right)^{j}+a_2\left(\dfrac{x}{y}\right)^{\frac{\sqrt{i}}{k}}+a_1=0$;

$$\text{x:} \quad a_{2n+2}\left(\frac{x}{y}\right)^{(n)j-\frac{\sqrt{i}}{k}} + a_{2n+1}\left(\frac{x}{y}\right)^{(n)j} + a_{2n}\left(\frac{x}{y}\right)^{(n-1)j-\frac{\sqrt{i}}{k}} + a_{2n-1}\left(\frac{x}{y}\right)^{(n-1)j} + ... + a_3\left(\frac{x}{y}\right)^{j} + a_2\left(\frac{x}{y}\right)^{-\frac{\sqrt{i}}{k}} + a_1 = 0 \; ;$$

$$\text{xi:} \quad a_{2n+2}\left(\frac{x}{y}\right)^{(n)(-j)+\frac{\sqrt{i}}{k}} + a_{2n+1}\left(\frac{x}{y}\right)^{(n)(-j)} + a_{2n}\left(\frac{x}{y}\right)^{(n-1)(-j)+\frac{\sqrt{i}}{k}} + a_{2n-1}\left(\frac{x}{y}\right)^{(n-1)(-j)} + ... + a_3\left(\frac{x}{y}\right)^{-j} + a_2\left(\frac{x}{y}\right)^{\frac{\sqrt{i}}{k}} + a_1 = 0 \; ;$$

$$\text{xii:} \quad a_{2n+2}\left(\frac{x}{y}\right)^{(n)(-j)-\frac{\sqrt{i}}{k}} + a_{2n+1}\left(\frac{x}{y}\right)^{(n)(-j)} + a_{2n}\left(\frac{x}{y}\right)^{(n-1)(-j)-\frac{\sqrt{i}}{k}} + a_{2n-1}\left(\frac{x}{y}\right)^{(n-1)(-j)} + ... + a_3\left(\frac{x}{y}\right)^{-j} + a_2\left(\frac{x}{y}\right)^{-\frac{\sqrt{i}}{k}} + a_1 = 0 \; ;$$

$$\text{I:} \quad a_{2n+2}\left(\frac{x}{y}\right)^{(n)\frac{i}{k}+j} + a_{2n+1}\left(\frac{x}{y}\right)^{(n)\frac{i}{k}} + a_{2n}\left(\frac{x}{y}\right)^{(n-1)\frac{i}{k}+j} + a_{2n-1}\left(\frac{x}{y}\right)^{(n-1)\frac{i}{k}} + ... + a_3\left(\frac{x}{y}\right)^{\frac{i}{k}} + a_2\left(\frac{x}{y}\right)^{j} + a_1 = 0 \; ;$$

$$\text{II:} \quad a_{2n+2}\left(\frac{x}{y}\right)^{(n)\frac{i}{k}-j} + a_{2n+1}\left(\frac{x}{y}\right)^{(n)\frac{i}{k}} + a_{2n}\left(\frac{x}{y}\right)^{(n-1)\frac{i}{k}-j} + a_{2n-1}\left(\frac{x}{y}\right)^{(n-1)\frac{i}{k}} + ... + a_3\left(\frac{x}{y}\right)^{\frac{i}{k}} + a_2\left(\frac{x}{y}\right)^{-j} + a_1 = 0 \; ;$$

$$\text{III:} \quad a_{2n+2}\left(\frac{x}{y}\right)^{(n)(-\frac{i}{k})+j} + a_{2n+1}\left(\frac{x}{y}\right)^{(n)(-\frac{i}{k})} + a_{2n}\left(\frac{x}{y}\right)^{(n-1)(-\frac{i}{k})+j} + a_{2n-1}\left(\frac{x}{y}\right)^{(n-1)(-\frac{i}{k})} + ... + a_3\left(\frac{x}{y}\right)^{-\frac{i}{k}} + a_2\left(\frac{x}{y}\right)^{j} + a_1 = 0 \; ;$$

$$\text{IV:} \quad a_{2n+2}\left(\frac{x}{y}\right)^{(n)(-\frac{i}{k})-j} + a_{2n+1}\left(\frac{x}{y}\right)^{(n)(-\frac{i}{k})} + a_{2n}\left(\frac{x}{y}\right)^{(n-1)(-\frac{i}{k})-j} + a_{2n-1}\left(\frac{x}{y}\right)^{(n-1)(-\frac{i}{k})} + ... + a_3\left(\frac{x}{y}\right)^{-\frac{i}{k}} + a_2\left(\frac{x}{y}\right)^{-j} + a_1 = 0 \; ;$$

$$\text{V:} \quad a_{2n+2}\left(\frac{x}{y}\right)^{(n)\sqrt{i}+j} + a_{2n+1}\left(\frac{x}{y}\right)^{(n)\sqrt{i}} + a_{2n}\left(\frac{x}{y}\right)^{(n-1)\sqrt{i}+j} + a_{2n-1}\left(\frac{x}{y}\right)^{(n-1)\sqrt{i}} + ... + a_3\left(\frac{x}{y}\right)^{\sqrt{i}} + a_2\left(\frac{x}{y}\right)^{j} + a_1 = 0 ;$$

$$\text{VI:} \quad a_{2n+2}\left(\frac{x}{y}\right)^{(n)\sqrt{i}-j} + a_{2n+1}\left(\frac{x}{y}\right)^{(n)\sqrt{i}} + a_{2n}\left(\frac{x}{y}\right)^{(n-1)\sqrt{i}-j} + a_{2n-1}\left(\frac{x}{y}\right)^{(n-1)\sqrt{i}} + ... + a_3\left(\frac{x}{y}\right)^{\sqrt{i}} + a_2\left(\frac{x}{y}\right)^{-j} + a_1 = 0 \; ;$$

$$\text{VII:} \quad a_{2n+2}\left(\frac{x}{y}\right)^{(n)(-\sqrt{i})+j} + a_{2n+1}\left(\frac{x}{y}\right)^{(n)(-\sqrt{i})} + a_{2n}\left(\frac{x}{y}\right)^{(n-1)(-\sqrt{i})+j} + a_{2n-1}\left(\frac{x}{y}\right)^{(n-1)(-\sqrt{i})} + ... + a_3\left(\frac{x}{y}\right)^{-\sqrt{i}} + a_2\left(\frac{x}{y}\right)^{j} + a_1 = 0 \; ;$$

$$\text{VIII:} \quad a_{2n+2}\left(\frac{x}{y}\right)^{(n)(-\sqrt{i})-j} + a_{2n+1}\left(\frac{x}{y}\right)^{(n)(-\sqrt{i})} + a_{2n}\left(\frac{x}{y}\right)^{(n-1)(-\sqrt{i})-j} + a_{2n-1}\left(\frac{x}{y}\right)^{(n-1)(-\sqrt{i})} + ... + a_3\left(\frac{x}{y}\right)^{-\sqrt{i}} + a_2\left(\frac{x}{y}\right)^{-j} + a_1 = 0 \; ;$$

$$\text{IX:} \quad a_{2n+2}\left(\frac{x}{y}\right)^{(n)\frac{\sqrt{i}}{k}+j} + a_{2n+1}\left(\frac{x}{y}\right)^{(n)\frac{\sqrt{i}}{k}} + a_{2n}\left(\frac{x}{y}\right)^{(n-1)\frac{\sqrt{i}}{k}+j} + a_{2n-1}\left(\frac{x}{y}\right)^{(n-1)\frac{\sqrt{i}}{k}} + ... + a_3\left(\frac{x}{y}\right)^{\frac{\sqrt{i}}{k}} + a_2\left(\frac{x}{y}\right)^{j} + a_1 = 0 \; ;$$

X: $a_{2n+2}\left(\dfrac{x}{y}\right)^{(n)\frac{\sqrt{i}}{k}-j} + a_{2n+1}\left(\dfrac{x}{y}\right)^{(n)\frac{\sqrt{i}}{k}} + a_{2n}\left(\dfrac{x}{y}\right)^{(n-1)\frac{\sqrt{i}}{k}-j} + a_{2n-1}\left(\dfrac{x}{y}\right)^{(n-1)\frac{\sqrt{i}}{k}} + ... + a_3\left(\dfrac{x}{y}\right)^{\frac{\sqrt{i}}{k}} + a_2\left(\dfrac{x}{y}\right)^{-j} + a_1 = 0$;

XI: $a_{2n+2}\left(\dfrac{x}{y}\right)^{(n)(-\frac{\sqrt{i}}{k})+j} + a_{2n+1}\left(\dfrac{x}{y}\right)^{(n)(-\frac{\sqrt{i}}{k})} + a_{2n}\left(\dfrac{x}{y}\right)^{(n-1)(-\frac{\sqrt{i}}{k})+j} + a_{2n-1}\left(\dfrac{x}{y}\right)^{(n-1)(-\frac{\sqrt{i}}{k})} + ... + a_3\left(\dfrac{x}{y}\right)^{-\frac{\sqrt{i}}{k}} + a_2\left(\dfrac{x}{y}\right)^{j} + a_1 = 0$;

XII: $a_{2n+2}\left(\dfrac{x}{y}\right)^{(n)(-\frac{\sqrt{i}}{k})-j} + a_{2n+1}\left(\dfrac{x}{y}\right)^{(n)(-\frac{\sqrt{i}}{k})} + a_{2n}\left(\dfrac{x}{y}\right)^{(n-1)(-\frac{\sqrt{i}}{k})-j} + a_{2n-1}\left(\dfrac{x}{y}\right)^{(n-1)(-\frac{\sqrt{i}}{k})} + ... + a_3\left(\dfrac{x}{y}\right)^{-\frac{\sqrt{i}}{k}} + a_2\left(\dfrac{x}{y}\right)^{-j} + a_1 = 0$;

$$n = 2,3,4,...,\infty; i = 2,3,4,...,\infty; k = 2,3,4,...,\infty; j = 2,3,4,...,\infty$$

(1) : A Composite-Multivariate Dinbakish-Funn Equation

$$a_4\left(\frac{x}{y}\right)^{2+\frac{2}{3}} + a_3\left(\frac{x}{y}\right)^{2} + a_2\left(\frac{x}{y}\right)^{\frac{2}{3}} + a_1 = 0 ,$$

One of the forms of this Composite-Multivariate Dinbakish-Funn Equation can be:

$$\left[a_4\left(\frac{x}{y}\right)-3\right]\left[\left(\frac{x}{y}\right)-4\right] = 0 ,$$

It expands:

$$a_4\left(\frac{x}{y}\right)^{2+\frac{2}{3}} - 4a_4\left(\frac{x}{y}\right)^{2} - 3\left(\frac{x}{y}\right)^{\frac{2}{3}} + 12 = 0 ,$$

And $a_3 = -4a_4; a_2 = -3; a_1 = 12$.

The Composite-Multivariate Dinbakish-Funn Equation follows the real number a_4 to be changed.

When $a_4 = 5$,it means that:

$$a_3 = -20 .$$

Now, the Composite-Multivariate Dinbakish-Funn Equation is:

$$5\left(\frac{x}{y}\right)^{2+\frac{2}{3}} - 20\left(\frac{x}{y}\right)^{2} - 3\left(\frac{x}{y}\right)^{\frac{2}{3}} + 12 = 0 \ .$$

In which it has two parts:

Part 1: $5\left(\frac{x}{y}\right)^{2} - 3 = 0$; Part 2: $\left(\frac{x}{y}\right)^{\frac{2}{3}} - 4 = 0$.

In the part 1:

$5\left(\frac{x}{y}\right)^{2} - 3 = 0$ that means: ① $\dfrac{x}{y} = \sqrt{\dfrac{3}{5}}$; and ② $\dfrac{x}{y} = -\sqrt{\dfrac{3}{5}}$.

Because $\dfrac{x}{y} = \sqrt{\dfrac{3}{5}}$, When $x = 1$, it means $y = \dfrac{1}{\sqrt{\dfrac{3}{5}}}$.

Also, because $\dfrac{x}{y} = -\sqrt{\dfrac{3}{5}}$, when $x = 2$, it means

$$y = -\dfrac{2}{\sqrt{\dfrac{3}{5}}} \ .$$

In the part 2:

$\left(\dfrac{x}{y}\right)^{\frac{2}{3}} - 4 = 0$ that means $\dfrac{x}{y} = 4^{\frac{3}{2}}$. When $x = -1$, it means $y = -\dfrac{1}{4^{\frac{3}{2}}}$.

(2) : A Composite-Multivariate Dinbakish-Funn Equation

$$a_4\left(\frac{x}{y}\right)^{\sqrt{3}+4} + a_3\left(\frac{x}{y}\right)^{\sqrt{3}} + a_2\left(\frac{x}{y}\right)^{4} + a_1 = 0$$

One of the forms of this Composite-Multivariate Dinbakish-Funn Equation can be:

$$\left[a_4\left(\frac{x}{y}\right)^{\sqrt{3}}-4\right]\left[\left(\frac{x}{y}\right)^4+2\right]=0 \ ,$$

It expands:

$$a_4\left(\frac{x}{y}\right)^{\sqrt{3}+4}+2a_4\left(\frac{x}{y}\right)^{\sqrt{3}}-4\left(\frac{x}{y}\right)^4-8=0 \ ,$$

And $a_3=2a_4; a_2=-4; a_1=-8$.

The Composite-Multivariate Dinbakish-Funn Equation follows the real number a_4 to be changed.

When $a_4=-1$, it means $a_3=-2$.

Now, the Composite-Multivariate Dinbakish-Funn Equation is:

$$-\left(\frac{x}{y}\right)^{\sqrt{3}+4}-2\left(\frac{x}{y}\right)^{\sqrt{3}}-4\left(\frac{x}{y}\right)^4-8=0 \ .$$

In which it has two parts: Part 1: $-\left(\frac{x}{y}\right)^{\sqrt{3}}-4=0$; Part 2: $\left(\frac{x}{y}\right)^4+2=0$.

In the part 1, $\frac{x}{y}=(-4)^{\frac{\sqrt{3}}{3}}$; when $x=\frac{1}{2}$, it means $y=\dfrac{1}{2(-4)^{\frac{\sqrt{3}}{3}}}$.

In the part 2, $\frac{x}{y}=(-2)^{\frac{1}{4}}$; when $x=-\frac{1}{2}$, it means $y=-\dfrac{1}{2(-2)^{\frac{1}{4}}}$.

(3) : A Composite-Multivariate Dinbakish-Funn Equation

$$a_4\left(\frac{x}{y}\right)^{5-\frac{\sqrt{3}}{4}}+a_3\left(\frac{x}{y}\right)^5+a_2\left(\frac{x}{y}\right)^{-\frac{\sqrt{3}}{4}}+a_1=0$$

One of the forms of this Composite-Multivariate Dinbakish-Funn Equation

$$\left[\left(\frac{x}{y}\right)^5 - 5\right]\left[a_4\left(\frac{x}{y}\right)^{-\frac{\sqrt{3}}{4}} + 3\right] = 0,$$

It expands:

$$a_4\left(\frac{x}{y}\right)^{5-\frac{\sqrt{3}}{4}} + 3\left(\frac{x}{y}\right)^5 - 5a_4\left(\frac{x}{y}\right)^{-\frac{\sqrt{3}}{4}} - 15 = 0, \text{ and } a_3 = 3; a_2 = -5a_4; a_1 = -15.$$

The Composite-Multivariate Dinbakish-Funn Equation follows the real number a_4 to be changed.

When $a_4 = 7$, it means that $a_2 = -35$.

Now, the Composite-Multivariate Dinbakish-Funn Equation is:

$$7\left(\frac{x}{y}\right)^{5-\frac{\sqrt{3}}{4}} + 3\left(\frac{x}{y}\right)^5 - 35\left(\frac{x}{y}\right)^{-\frac{\sqrt{3}}{4}} - 15 = 0.$$

In which it has two parts: Part 1: $\left(\frac{x}{y}\right)^5 - 5 = 0$; Part 2: $7\left(\frac{x}{y}\right)^{-\frac{\sqrt{3}}{4}} + 3 = 0$.

In the part 1: $\dfrac{x}{y} = 5^{\frac{1}{5}}$, when $x = \dfrac{1}{3}$, it means $y = \dfrac{1}{3 \times 5^{\frac{1}{5}}}$.

In the part 2: $\dfrac{x}{y} = \left(-\dfrac{3}{7}\right)^{-\frac{4\sqrt{3}}{3}}$, when $x = -\dfrac{1}{3}$, it means $y = -\dfrac{1}{3\left(-\dfrac{3}{7}\right)^{-\frac{4\sqrt{3}}{3}}}$.

(4) : A Composite-Multivariate Dinbakish-Funn Equation

$$a_4\left(\frac{x}{y}\right)^{6+2\sqrt{3}} + a_3\left(\frac{x}{y}\right)^6 + a_2\left(\frac{x}{y}\right)^{2\sqrt{3}} + a_1 = 0$$

One of the forms of this Composite-Multivariate Dinbakish-Funn Equation can be:

$$\left[\left(\frac{x}{y}\right)^6 - a_4\right]\left[a_4\left(\frac{x}{y}\right)^{2\sqrt{3}} + 1\right] = 0 \ ,$$

It expands:

$$a_4\left(\frac{x}{y}\right)^{6+2\sqrt{3}} + \left(\frac{x}{y}\right)^6 - a_4^2\left(\frac{x}{y}\right)^{2\sqrt{3}} - a_4 = 0 \ , \text{ and } a_3 = 1; a_2 = -a_4^2; a_1 = -a_4.$$

The Composite-Multivariate Dinbakish-Funn Equation follows the real number a_4 to be changed.

When $a_4 = -4$, it means that $a_2 = -16; a_1 = 4$.

Now, the Composite-Multivariate Dinbakish-Funn Equation is:

$$-4\left(\frac{x}{y}\right)^{6+2\sqrt{3}} + \left(\frac{x}{y}\right)^6 - 16\left(\frac{x}{y}\right)^{2\sqrt{3}} + 4 = 0 \ .$$

In which it has two parts: Part 1: $\left(\frac{x}{y}\right)^6 + 4 = 0$; Part 2: $-4\left(\frac{x}{y}\right)^{2\sqrt{3}} + 1 = 0$.

In the part 1: $\dfrac{x}{y} = (-4)^{\frac{1}{6}}$, when $x = \dfrac{1}{4}$, it means $y = \dfrac{1}{4(-4)^{\frac{1}{6}}}$.

In the part 2: $\dfrac{x}{y} = \left(\dfrac{1}{4}\right)^{\frac{\sqrt{3}}{6}}$, when $x = -\dfrac{1}{4}$, it means $y = -\dfrac{1}{4\left(\dfrac{1}{4}\right)^{\frac{\sqrt{3}}{6}}}$.

(5) : A Composite-Multivariate Dinbakish-Funn Equation

$$a_4\left(\frac{x}{y}\right)^{\frac{3}{4}-7} + a_3\left(\frac{x}{y}\right)^{\frac{3}{4}} + a_2\left(\frac{x}{y}\right)^{-7} + a_1 = 0$$

One of the forms of this Composite-Multivariate Dinbakish Equation can be:

$$\left[\left(\frac{x}{y}\right)^{\frac{3}{4}}+a_4\right]\left[a_4\left(\frac{x}{y}\right)^{-7}+3\right]=0 \ ,$$

It expands:

$$a_4\left(\frac{x}{y}\right)^{\frac{3}{4}-7}+3\left(\frac{x}{y}\right)^{\frac{3}{4}}+a_4^2\left(\frac{x}{y}\right)^{-7}+3a_4=0 \ , \text{ and } a_3=3; a_2=a_4^2; a_1=3a_4 \ .$$

The Composite-Multivariate Dinbakish-Funn Equation follows the real number a_4 to be changed.

When $a_4=-3$, it means that $a_2=9; a_1=-9$.

Now, the Composite-Multivariate Dinbakish-Funn Equation is:

$$-3\left(\frac{x}{y}\right)^{\frac{3}{4}-7}+3\left(\frac{x}{y}\right)^{\frac{3}{4}}+9\left(\frac{x}{y}\right)^{-7}-9=0 \ .$$

In which it has two parts: Part 1: $\left(\frac{x}{y}\right)^{\frac{3}{4}}-3=0$; Part 2: $-3\left(\frac{x}{y}\right)^{-7}+3=0$.

In the part 1: $\dfrac{x}{y}=3^{\frac{4}{3}}$, when $x=\dfrac{1}{7}$, it means $y=\dfrac{1}{7\times 3^{\frac{4}{3}}}$.

In the part 2: $\dfrac{x}{y}=1$, that it means $x=y$.

(6) : The other Composite-Multivariate Dinbakish-Funn Equations

[1]: $a_{2n+2}\left(\dfrac{x}{yz}\right)^{(n)j+\frac{i}{k}}+a_{2n+1}\left(\dfrac{x}{yz}\right)^{(n)j}+a_{2n}\left(\dfrac{x}{yz}\right)^{(n-1)j+\frac{i}{k}}+a_{2n-1}\left(\dfrac{x}{yz}\right)^{(n-1)j}+...+a_3\left(\dfrac{x}{yz}\right)^{j}+a_2\left(\dfrac{x}{yz}\right)^{\frac{i}{k}}+a_1=0$;

[2]: $a_{2n+2}\left(\dfrac{x}{yz}\right)^{(n)j-\frac{i}{k}}+a_{2n+1}\left(\dfrac{x}{yz}\right)^{(n)j}+a_{2n}\left(\dfrac{x}{yz}\right)^{(n-1)j-\frac{i}{k}}+a_{2n-1}\left(\dfrac{x}{yz}\right)^{(n-1)j}+...+a_3\left(\dfrac{x}{yz}\right)^{j}+a_2\left(\dfrac{x}{yz}\right)^{-\frac{i}{k}}+a_1=0$;

$[3]$: $a_{2n+2}\left(\dfrac{x}{yz}\right)^{(n)(-j)+\frac{i}{k}} + a_{2n+1}\left(\dfrac{x}{yz}\right)^{(n)(-j)} + a_{2n}\left(\dfrac{x}{yz}\right)^{(n-1)(-j)+\frac{i}{k}} + a_{2n-1}\left(\dfrac{x}{yz}\right)^{(n-1)(-j)} + ... + a_3\left(\dfrac{x}{yz}\right)^{-j} + a_2\left(\dfrac{x}{yz}\right)^{\frac{i}{k}} + a_1 = 0$;

$[4]$: $a_{2n+2}\left(\dfrac{x}{yz}\right)^{(n)(-j)-\frac{i}{k}} + a_{2n+1}\left(\dfrac{x}{yz}\right)^{(n)(-j)} + a_{2n}\left(\dfrac{x}{yz}\right)^{(n-1)(-j)-\frac{i}{k}} + a_{2n-1}\left(\dfrac{x}{yz}\right)^{(n-1)(-j)} + ... + a_3\left(\dfrac{x}{yz}\right)^{-j} + a_2\left(\dfrac{x}{yz}\right)^{-\frac{i}{k}} + a_1 = 0$;

$[5]$: $a_{2n+2}\left(\dfrac{x}{yz}\right)^{(n)j+\sqrt{i}} + a_{2n+1}\left(\dfrac{x}{yz}\right)^{(n)j} + a_{2n}\left(\dfrac{x}{yz}\right)^{(n-1)j+\sqrt{i}} + a_{2n-1}\left(\dfrac{x}{yz}\right)^{(n-1)j} + ... + a_3\left(\dfrac{x}{yz}\right)^{j} + a_2\left(\dfrac{x}{yz}\right)^{\sqrt{i}} + a_1 = 0$;

$[6]$: $a_{2n+2}\left(\dfrac{x}{yz}\right)^{(n)j-\sqrt{i}} + a_{2n+1}\left(\dfrac{x}{yz}\right)^{(n)j} + a_{2n}\left(\dfrac{x}{yz}\right)^{(n-1)j-\sqrt{i}} + a_{2n-1}\left(\dfrac{x}{yz}\right)^{(n-1)j} + ... + a_3\left(\dfrac{x}{yz}\right)^{j} + a_2\left(\dfrac{x}{yz}\right)^{-\sqrt{i}} + a_1 = 0$;

$[7]$: $a_{2n+2}\left(\dfrac{x}{yz}\right)^{(n)(-j)+\sqrt{i}} + a_{2n+1}\left(\dfrac{x}{yz}\right)^{(n)(-j)} + a_{2n}\left(\dfrac{x}{yz}\right)^{(n-1)(-j)+\sqrt{i}} + a_{2n-1}\left(\dfrac{x}{yz}\right)^{(n-1)(-j)} + ... + a_3\left(\dfrac{x}{yz}\right)^{-j} + a_2\left(\dfrac{x}{yz}\right)^{\sqrt{i}} + a_1 = 0$;

$[8]$: $a_{2n+2}\left(\dfrac{x}{yz}\right)^{(n)(-j)-\sqrt{i}} + a_{2n+1}\left(\dfrac{x}{yz}\right)^{(n)(-j)} + a_{2n}\left(\dfrac{x}{yz}\right)^{(n-1)(-j)-\sqrt{i}} + a_{2n-1}\left(\dfrac{x}{yz}\right)^{(n-1)(-j)} + ... + a_3\left(\dfrac{x}{yz}\right)^{-j} + a_2\left(\dfrac{x}{yz}\right)^{-\sqrt{i}} + a_1 = 0$;

$[9]$: $a_{2n+2}\left(\dfrac{x}{yz}\right)^{(n)j+\frac{\sqrt{i}}{k}} + a_{2n+1}\left(\dfrac{x}{yz}\right)^{(n)j} + a_{2n}\left(\dfrac{x}{yz}\right)^{(n-1)j+\frac{\sqrt{i}}{k}} + a_{2n-1}\left(\dfrac{x}{yz}\right)^{(n-1)j} + ... + a_3\left(\dfrac{x}{yz}\right)^{j} + a_2\left(\dfrac{x}{yz}\right)^{\frac{\sqrt{i}}{k}} + a_1 = 0$;

$[10]$: $a_{2n+2}\left(\dfrac{x}{yz}\right)^{(n)j-\frac{\sqrt{i}}{k}} + a_{2n+1}\left(\dfrac{x}{yz}\right)^{(n)j} + a_{2n}\left(\dfrac{x}{yz}\right)^{(n-1)j-\frac{\sqrt{i}}{k}} + a_{2n-1}\left(\dfrac{x}{yz}\right)^{(n-1)j} + ... + a_3\left(\dfrac{x}{yz}\right)^{j} + a_2\left(\dfrac{x}{yz}\right)^{-\frac{\sqrt{i}}{k}} + a_1 = 0$;

$[11]$: $a_{2n+2}\left(\dfrac{x}{yz}\right)^{(n)(-j)+\frac{\sqrt{i}}{k}} + a_{2n+1}\left(\dfrac{x}{yz}\right)^{(n)(-j)} + a_{2n}\left(\dfrac{x}{yz}\right)^{(n-1)(-j)+\frac{\sqrt{i}}{k}} + a_{2n-1}\left(\dfrac{x}{yz}\right)^{(n-1)(-j)} + ... + a_3\left(\dfrac{x}{yz}\right)^{-j} + a_2\left(\dfrac{x}{yz}\right)^{\frac{\sqrt{i}}{k}} + a_1 = 0$;

$[12]$: $a_{2n+2}\left(\dfrac{x}{yz}\right)^{(n)(-j)-\frac{\sqrt{i}}{k}} + a_{2n+1}\left(\dfrac{x}{yz}\right)^{(n)(-j)} + a_{2n}\left(\dfrac{x}{yz}\right)^{(n-1)(-j)-\frac{\sqrt{i}}{k}} + a_{2n-1}\left(\dfrac{x}{yz}\right)^{(n-1)(-j)} + ... + a_3\left(\dfrac{x}{yz}\right)^{-j} + a_2\left(\dfrac{x}{yz}\right)^{-\frac{\sqrt{i}}{k}} + a_1 = 0$;

$[13]$: $a_{2n+2}\left(\dfrac{x}{yz}\right)^{(n)\frac{i}{k}+j} + a_{2n+1}\left(\dfrac{x}{yz}\right)^{(n)\frac{i}{k}} + a_{2n}\left(\dfrac{x}{yz}\right)^{(n-1)\frac{i}{k}+j} + a_{2n-1}\left(\dfrac{x}{yz}\right)^{(n-1)\frac{i}{k}} + ... + a_3\left(\dfrac{x}{yz}\right)^{\frac{i}{k}} + a_2\left(\dfrac{x}{yz}\right)^{j} + a_1 = 0$;

$[14]$: $a_{2n+2}\left(\dfrac{x}{yz}\right)^{(n)\frac{i}{k}-j} + a_{2n+1}\left(\dfrac{x}{yz}\right)^{(n)\frac{i}{k}} + a_{2n}\left(\dfrac{x}{yz}\right)^{(n-1)\frac{i}{k}-j} + a_{2n-1}\left(\dfrac{x}{yz}\right)^{(n-1)\frac{i}{k}} + ... + a_3\left(\dfrac{x}{yz}\right)^{\frac{i}{k}} + a_2\left(\dfrac{x}{yz}\right)^{-j} + a_1 = 0$;

$$[15]:\ a_{2n+2}\left(\frac{x}{yz}\right)^{(n)\left(-\frac{i}{k}\right)+j} + a_{2n+1}\left(\frac{x}{yz}\right)^{(n)\left(-\frac{i}{k}\right)} + a_{2n}\left(\frac{x}{yz}\right)^{(n-1)\left(-\frac{i}{k}\right)+j} + a_{2n-1}\left(\frac{x}{yz}\right)^{(n-1)\left(-\frac{i}{k}\right)} + ... + a_3\left(\frac{x}{yz}\right)^{-\frac{i}{k}} + a_2\left(\frac{x}{yz}\right)^{j} + a_1 = 0\ ;$$

$$[16]:\ a_{2n+2}\left(\frac{x}{yz}\right)^{(n)\left(-\frac{i}{k}\right)-j} + a_{2n+1}\left(\frac{x}{yz}\right)^{(n)\left(-\frac{i}{k}\right)} + a_{2n}\left(\frac{x}{yz}\right)^{(n-1)\left(-\frac{i}{k}\right)-j} + a_{2n-1}\left(\frac{x}{yz}\right)^{(n-1)\left(-\frac{i}{k}\right)} + ... + a_3\left(\frac{x}{yz}\right)^{-\frac{i}{k}} + a_2\left(\frac{x}{yz}\right)^{-j} + a_1 = 0\ ;$$

$$[17]:\ a_{2n+2}\left(\frac{x}{yz}\right)^{(n)\frac{\sqrt{i}}{k}+j} + a_{2n+1}\left(\frac{x}{yz}\right)^{(n)\frac{\sqrt{i}}{k}} + a_{2n}\left(\frac{x}{yz}\right)^{(n-1)\frac{\sqrt{i}}{k}+j} + a_{2n-1}\left(\frac{x}{yz}\right)^{(n-1)\frac{\sqrt{i}}{k}} + ... + a_3\left(\frac{x}{yz}\right)^{\frac{\sqrt{i}}{k}} + a_2\left(\frac{x}{yz}\right)^{j} + a_1 = 0\ ;$$

$$[18]:\ a_{2n+2}\left(\frac{x}{yz}\right)^{(n)\frac{\sqrt{i}}{k}-j} + a_{2n+1}\left(\frac{x}{yz}\right)^{(n)\frac{\sqrt{i}}{k}} + a_{2n}\left(\frac{x}{yz}\right)^{(n-1)\frac{\sqrt{i}}{k}-j} + a_{2n-1}\left(\frac{x}{yz}\right)^{(n-1)\frac{\sqrt{i}}{k}} + ... + a_3\left(\frac{x}{yz}\right)^{\frac{\sqrt{i}}{k}} + a_2\left(\frac{x}{yz}\right)^{-j} + a_1 = 0\ ;$$

$$[19]:\ a_{2n+2}\left(\frac{x}{yz}\right)^{(n)\left(-\frac{\sqrt{i}}{k}\right)+j} + a_{2n+1}\left(\frac{x}{yz}\right)^{(n)\left(-\frac{\sqrt{i}}{k}\right)} + a_{2n}\left(\frac{x}{yz}\right)^{(n-1)\left(-\frac{\sqrt{i}}{k}\right)+j} + a_{2n-1}\left(\frac{x}{yz}\right)^{(n-1)\left(-\frac{\sqrt{i}}{k}\right)} + ... + a_3\left(\frac{x}{yz}\right)^{-\frac{\sqrt{i}}{k}} + a_2\left(\frac{x}{yz}\right)^{j} + a_1 = 0\ ;$$

$$[20]:\ a_{2n+2}\left(\frac{x}{yz}\right)^{(n)\left(-\frac{\sqrt{i}}{k}\right)-j} + a_{2n+1}\left(\frac{x}{yz}\right)^{(n)\left(-\frac{\sqrt{i}}{k}\right)} + a_{2n}\left(\frac{x}{yz}\right)^{(n-1)\left(-\frac{\sqrt{i}}{k}\right)-j} + a_{2n-1}\left(\frac{x}{yz}\right)^{(n-1)\left(-\frac{\sqrt{i}}{k}\right)} + ... + a_3\left(\frac{x}{yz}\right)^{-\frac{\sqrt{i}}{k}} + a_2\left(\frac{x}{yz}\right)^{-j} + a_1 = 0\ ;$$

$$[21]:\ a_{2n+2}\left(\frac{x}{yz}\right)^{(n)\sqrt{i}+j} + a_{2n+1}\left(\frac{x}{yz}\right)^{(n)\sqrt{i}} + a_{2n}\left(\frac{x}{yz}\right)^{(n-1)\sqrt{i}+j} + a_{2n-1}\left(\frac{x}{yz}\right)^{(n-1)\sqrt{i}} + ... + a_3\left(\frac{x}{yz}\right)^{\sqrt{i}} + a_2\left(\frac{x}{yz}\right)^{j} + a_1 = 0\ ;$$

$$[22]:\ a_{2n+2}\left(\frac{x}{yz}\right)^{(n)\sqrt{i}-j} + a_{2n+1}\left(\frac{x}{yz}\right)^{(n)\sqrt{i}} + a_{2n}\left(\frac{x}{yz}\right)^{(n-1)\sqrt{i}-j} + a_{2n-1}\left(\frac{x}{yz}\right)^{(n-1)\sqrt{i}} + ... + a_3\left(\frac{x}{yz}\right)^{\sqrt{i}} + a_2\left(\frac{x}{yz}\right)^{-j} + a_1 = 0\ ;$$

$$[23]:\ a_{2n+2}\left(\frac{x}{yz}\right)^{(n)(-\sqrt{i})+j} + a_{2n+1}\left(\frac{x}{yz}\right)^{(n)(-\sqrt{i})} + a_{2n}\left(\frac{x}{yz}\right)^{(n-1)(-\sqrt{i})+j} + a_{2n-1}\left(\frac{x}{yz}\right)^{(n-1)(-\sqrt{i})} + ... + a_3\left(\frac{x}{yz}\right)^{-\sqrt{i}} + a_2\left(\frac{x}{yz}\right)^{j} + a_1 = 0\ ;$$

$$[24]:\ a_{2n+2}\left(\frac{x}{yz}\right)^{(n)(-\sqrt{i})-j} + a_{2n+1}\left(\frac{x}{yz}\right)^{(n)(-\sqrt{i})} + a_{2n}\left(\frac{x}{yz}\right)^{(n-1)(-\sqrt{i})-j} + a_{2n-1}\left(\frac{x}{yz}\right)^{(n-1)(\sqrt{i})} + ... + a_3\left(\frac{x}{yz}\right)^{-\sqrt{i}} + a_2\left(\frac{x}{yz}\right)^{-j} + a_1 = 0\ ;$$

$$n = 2,3,4,...,\infty; i = 2,3,4,...,\infty; k = 2,3,4,...,\infty; j = 2,3,4,...,\infty; y = a_2 x + a_1$$

About the complex Composite-Multivariate Dinbakish-Funn Equations, I have no time to show them in the Volume 2. In fact, they have other things in which they are always great.

Part 3

The Dinbakish Equations and their Inverse Functions

(1) : A Dinbakish Equation Function $f(x) = y = x^{\frac{1}{2}} + b$, $b \in R; b \neq 0;$

Now, looking at $y = x^{\frac{1}{2}} + b$, it means $x^{\frac{1}{2}} = y - b$, $x = (y-b)^2 = y^2 - 2by + b^2$,

And its inverse function is $f^{-1}(x) = x^2 - 2bx + b^2$, $b \in R; b \neq 0$.

(2) : A Dinbakish Equation Function $f(x) = y = kx^{\frac{1}{2}} + b$, $k \in R; b \in R; k \neq 0; b \neq 0$.

Now, looking at $y = kx^{\frac{1}{2}} + b$, it means $x^{\frac{1}{2}} = \dfrac{y-b}{k}$, $x = \dfrac{1}{k^2} y^2 - \dfrac{2b}{k^2} y + \dfrac{b^2}{k^2}$.

And its inverse function is $f^{-1}(x) = \dfrac{1}{k^2} x^2 - \dfrac{2b}{k^2} x + \dfrac{b^2}{k^2}$, $k \in R; b \in R; k \neq 0; b \neq 0$.

I: From (1) to (2) , about the inverse function of a Dinbakish Equation, it has a general form:

$$f(x) = kx^{\frac{1}{n}} + b , \ k \in R; b \in R; k \neq 0; b \neq 0; n = 1, 2, 3, ..., \infty ;$$

Its inverse function is:

$$f^{-1}(x) = \frac{1}{k^n} c_n^0 x^n + \frac{1}{k^n} c_n^1 x^{n-1} (-b) + \frac{1}{k^n} c_n^2 x^{n-2} (-b)^2 + ... + \frac{1}{k^n} c_n^{n-1} x (-b)^{n-1} + \frac{1}{k^n} c_n^n (-b)^n ;$$

They are about straight line $x = y$.

(3) : A Dinbakish Equation Function $f(x) = y = kx^{\frac{2}{3}} + b, k \in R; b \in R; k \neq 0; b \neq 0$.

Now, looking at $y = kx^{\frac{2}{3}} + b$, it means:

$$x = \left(\frac{y-b}{k} \right)^{\frac{3}{2}} = \frac{1}{k^{\frac{3}{2}}} (y-b)^{\frac{3}{2}} = \frac{1}{k^{\frac{3}{2}}} (y-b)^{\frac{1}{2}} (y-b)^{\frac{1}{2}} (y-b)^{\frac{1}{2}} = \frac{1}{k^{\frac{3}{2}}} y (y-b)^{\frac{1}{2}} - \frac{1}{k^{\frac{3}{2}}} b (y-b)^{\frac{1}{2}} ,$$

And its inverse function is: $f^{-1}(x) = \dfrac{1}{k^{\frac{3}{2}}} x(x-b)^{\frac{1}{2}} - \dfrac{1}{k^{\frac{3}{2}}} b(x-b)^{\frac{1}{2}}$.

(4) : A Dinbakish Equation Function $f(x) = y = kx^{\frac{2}{5}} + b, k \in R; b \in R; k \neq 0; b \neq 0$.

Now, looking at $y = kx^{\frac{2}{5}} + b$, it means:

$$x = \left(\dfrac{y-b}{k}\right)^{\frac{5}{2}} = \dfrac{1}{k^{\frac{5}{2}}}(y-b)^{\frac{5}{2}} = \dfrac{1}{k^{\frac{5}{2}}}\left(y^2 - 2by + b^2\right)(y-b)^{\frac{1}{2}} = \dfrac{1}{k^{\frac{5}{2}}}y^2(y-b)^{\frac{1}{2}} - \dfrac{1}{k^{\frac{5}{2}}}2by(y-b)^{\frac{1}{2}} + \dfrac{1}{k^{\frac{5}{2}}}b^2(y-b)^{\frac{1}{2}} .$$

And its inverse function is: $f^{-1}(x) = \dfrac{1}{k^{\frac{5}{2}}}x^2(x-b)^{\frac{1}{2}} - \dfrac{1}{k^{\frac{5}{2}}}2bx(x-b)^{\frac{1}{2}} + \dfrac{1}{k^{\frac{5}{2}}}b^2(x-b)^{\frac{1}{2}}$.

II: From (3) to (4) , about the inverse function of a Dinbakish Equation function, it has a general form.

$$f(x) = kx^{\frac{2}{2n+1}} + b, \ \ k \in R; b \in R; k \neq 0; b \neq 0; n = 1,2,3,...,\infty ;$$

Its inverse function is:

$$f^{-1}(x) = \dfrac{1}{k^{\frac{2n+1}{2}}}c_n^0 x^n (x-b)^{\frac{1}{2}} + \dfrac{1}{k^{\frac{2n+1}{2}}}c_n^1 x^{n-1}(-b)(x-b)^{\frac{1}{2}} + \dfrac{1}{k^{\frac{2n+1}{2}}}c_n^2 x^{n-2}(-b)^2(x-b)^{\frac{1}{2}} + ... + \dfrac{1}{k^{\frac{2n+1}{2}}}c_n^{n-1}x(-b)^{n-1}(x-b)^{\frac{1}{2}} + \dfrac{1}{k^{\frac{2n+1}{2}}}c_n^n(-b)^n(x-b)^{\frac{1}{2}} ;$$

They are about the straight line $x = y$.

(5) : A Dinbakish Equation Function $f(x) = y = kx^{\frac{3}{4}} + b, k \in R; b \in R; k \neq 0; b \neq 0$.

Now, looking at $y = kx^{\frac{3}{4}} + b$, it means:

$$x = \left(\dfrac{y-b}{k}\right)^{\frac{4}{3}} = \dfrac{1}{k^{\frac{4}{3}}}(y-b)^{\frac{4}{3}} = \dfrac{1}{k^{\frac{4}{3}}}y(y-b)^{\frac{1}{3}} - \dfrac{1}{k^{\frac{4}{3}}}b(y-b)^{\frac{1}{3}} .$$

And its inverse function is $f^{-1}(x) = \dfrac{1}{k^{\frac{4}{3}}}x(x-b)^{\frac{1}{3}} - \dfrac{1}{k^{\frac{4}{3}}}b(x-b)^{\frac{1}{3}}$.

(6) : A Dinbakish Equation Function $f(x)=y=kx^{\frac{3}{5}}+b, k\in R; b\in R; k\neq 0; b\neq 0$,

Now, looking at $y=kx^{\frac{3}{5}}+b$, it means:

$$x=\left(\frac{y-b}{k}\right)^{\frac{5}{3}}=\frac{1}{k^{\frac{5}{3}}}(y-b)^{\frac{5}{3}}=\frac{1}{k^{\frac{5}{3}}}y(y-b)^{\frac{2}{3}}-\frac{1}{k^{\frac{5}{3}}}b(y-b)^{\frac{2}{3}} ,$$

And its inverse function is $f^{-1}(x)=\frac{1}{k^{\frac{5}{3}}}x(x-b)^{\frac{2}{3}}-\frac{1}{k^{\frac{5}{3}}}b(x-b)^{\frac{2}{3}}$.

III: From (5) to (6) , about the inverse function of a Dinbakish Equation Function, its general forms have two parts.

Part 1: $f(x)=kx^{\frac{3}{3n+1}}+b; k\in R; b\in R; k\neq 0; b\neq 0; n=1,2,3,...,\infty$.

Its inverse function is:

$$f^{-1}(x)=\frac{1}{k^{\frac{3n+1}{3}}}c_n^0 x^n(x-b)^{\frac{1}{3}}+\frac{1}{k^{\frac{3n+1}{3}}}c_n^1 x^{n-1}(-b)(x-b)^{\frac{1}{3}}+\frac{1}{k^{\frac{3n+1}{3}}}c_n^2 x^{n-2}(-b)^2(x-b)^{\frac{1}{3}}+...+\frac{1}{k^{\frac{3n+1}{3}}}c_n^{n-1}x(-b)^{n-1}(x-b)^{\frac{1}{3}}+\frac{1}{k^{\frac{3n+1}{3}}}c_n^n(-b)^n(x-b)^{\frac{1}{3}} .$$

Part 2: $f(x)=kx^{\frac{3}{3n+2}}+b; k\in R; b\in R; k\neq 0; b\neq 0; n=1,2,3,...,\infty$;

Its inverse function is:

$$f^{-1}(x)=\frac{1}{k^{\frac{3n+2}{3}}}c_n^0 x^n(x-b)^{\frac{2}{3}}+\frac{1}{k^{\frac{3n+2}{3}}}c_n^1 x^{n-1}(-b)(x-b)^{\frac{2}{3}}+\frac{1}{k^{\frac{3n+2}{3}}}c_n^2 x^{n-2}(-b)^2(x-b)^{\frac{2}{3}}+...+\frac{1}{k^{\frac{3n+2}{3}}}c_n^{n-1}x(-b)^{n-1}(x-b)^{\frac{2}{3}}+\frac{1}{k^{\frac{3n+2}{3}}}c_n^n(-b)^n(x-b)^{\frac{2}{3}} .$$

They are about straight line $x=y$.

(7) : A Dinbakish Equation Function $f(x)=y=kx^{\frac{4}{5}}+b, k\in R; b\in R; k\neq 0; b\neq 0$;

Now, looking at $y=kx^{\frac{4}{5}}+b$, it means:

$$x=\left(\frac{y-b}{k}\right)^{\frac{5}{4}}=\frac{1}{k^{\frac{5}{4}}}(y-b)^{\frac{5}{4}}=\frac{1}{k^{\frac{5}{4}}}y(y-b)^{\frac{1}{4}}-\frac{1}{k^{\frac{5}{4}}}b(y-b)^{\frac{1}{4}} .$$

Its inverse function is: $f^{-1}(x) = \dfrac{1}{k^{\frac{5}{4}}} x(x-b)^{\frac{1}{4}} - \dfrac{1}{k^{\frac{5}{4}}} b(x-b)^{\frac{1}{4}}$.

(8) : A Dinbakish Equation Function $f(x) = y = kx^{\frac{4}{6}} + b, k \in R; b \in R; k \neq 0; b \neq 0$.

Now, looking at $y = kx^{\frac{4}{6}} + b$, it means:

$$x = \left(\dfrac{y-b}{k}\right)^{\frac{6}{4}} = \dfrac{1}{k^{\frac{6}{4}}}(y-b)^{\frac{6}{4}} = \dfrac{1}{k^{\frac{6}{4}}} y(y-b)^{\frac{2}{4}} - \dfrac{1}{k^{\frac{6}{4}}} b(y-b)^{\frac{2}{4}} .$$

Its inverse function is: $f^{-1}(x) = \dfrac{1}{k^{\frac{6}{4}}} x(x-b)^{\frac{2}{4}} - \dfrac{1}{k^{\frac{6}{4}}} b(x-b)^{\frac{2}{4}}$.

IV: From (7) to (8) , about the inverse function of a Dinbakish Equation Function, its general forms have three parts.

Part 1: $f(x) = kx^{\frac{4}{4n+1}} + b; k \in R; b \in R; k \neq 0; b \neq 0; n = 1,2,3,...,\infty$.

Its inverse function is:

$$f^{-1}(x) = \dfrac{1}{k^{\frac{4n+1}{4}}} c_n^0 x^n (x-b)^{\frac{1}{4}} + \dfrac{1}{k^{\frac{4n+1}{4}}} c_n^1 x^{n-1}(-b)(x-b)^{\frac{1}{4}} + \dfrac{1}{k^{\frac{4n+1}{4}}} c_n^2 x^{n-2}(-b)^2(x-b)^{\frac{1}{4}} + ... + \dfrac{1}{k^{\frac{4n+1}{4}}} c_n^{n-1} x(-b)^{n-1}(x-b)^{\frac{1}{4}} + \dfrac{1}{k^{\frac{4n+1}{4}}} c_n^n (-b)^n (x-b)^{\frac{1}{4}} .$$

Part 2: $f(x) = kx^{\frac{4}{4n+2}} + b; k \in R; b \in R; k \neq 0; b \neq 0; n = 1,2,3,...,\infty$.

Its inverse function is:

$$f^{-1}(x) = \dfrac{1}{k^{\frac{4n+2}{4}}} c_n^0 x^n (x-b)^{\frac{2}{4}} + \dfrac{1}{k^{\frac{4n+2}{4}}} c_n^1 x^{n-1}(-b)(x-b)^{\frac{2}{4}} + \dfrac{1}{k^{\frac{4n+2}{4}}} c_n^2 x^{n-2}(-b)^2(x-b)^{\frac{2}{4}} + ... + \dfrac{1}{k^{\frac{4n+2}{4}}} c_n^{n-1} x(-b)^{n-1}(x-b)^{\frac{2}{4}} + \dfrac{1}{k^{\frac{4n+2}{4}}} c_n^n (-b)^n (x-b)^{\frac{2}{4}} .$$

Part 3: $f(x) = kx^{\frac{4}{4n+3}} + b; k \in R; b \in R; k \neq 0; b \neq 0; n = 1,2,3,...,\infty$.

Its inverse function is:

$$f^{-1}(x) = \dfrac{1}{k^{\frac{4n+3}{4}}} c_n^0 x^n (x-b)^{\frac{3}{4}} + \dfrac{1}{k^{\frac{4n+3}{4}}} c_n^1 x^{n-1}(-b)(x-b)^{\frac{3}{4}} + \dfrac{1}{k^{\frac{4n+3}{4}}} c_n^2 x^{n-2}(-b)^2(x-b)^{\frac{3}{4}} + ... + \dfrac{1}{k^{\frac{4n+3}{4}}} c_n^{n-1} x(-b)^{n-1}(x-b)^{\frac{3}{4}} + \dfrac{1}{k^{\frac{4n+3}{4}}} c_n^n (-b)^n (x-b)^{\frac{3}{4}} .$$

They are about straight line $x = y$.

V: And from (2) to (8) for making a section, about the inverse function of a Dinbakish Equation Function, its general forms have $m-1$ parts. $m = 2, 3, 4, ..., \infty$.

Part 1: $f(x) = kx^{\frac{m}{mn+1}} + b; a \in R; b \in R; a \neq 0; b \neq 0; m = 2, 3, 4, ..., \infty; n = 1, 2, 3, ..., \infty$.

Its inverse function is:

$$f^{-1}(x) = \frac{1}{k^{\frac{mn+1}{m}}} c_n^0 x^n (x-b)^{\frac{1}{m}} + \frac{1}{k^{\frac{mn+1}{m}}} c_n^1 x^{n-1} (-b)(x-b)^{\frac{1}{m}} + \frac{1}{k^{\frac{mn+1}{m}}} c_n^2 x^{n-2} (-b)^2 (x-b)^{\frac{1}{m}} + ... + \frac{1}{k^{\frac{mn+1}{m}}} c_n^{n-1} x (-b)^{n-1} (x-b)^{\frac{1}{m}} + \frac{1}{k^{\frac{mn+1}{m}}} c_n^n (-b)^n (x-b)^{\frac{1}{m}}.$$

Part 2: $f(x) = kx^{\frac{m}{mn+2}} + b; a \in R; b \in R; a \neq 0; b \neq 0; m = 2, 3, 4, ..., \infty; n = 1, 2, 3, ..., \infty$.

Its inverse function is:

$$f^{-1}(x) = \frac{1}{k^{\frac{mn+2}{m}}} c_n^0 x^n (x-b)^{\frac{2}{m}} + \frac{1}{k^{\frac{mn+2}{m}}} c_n^1 x^{n-1} (-b)(x-b)^{\frac{2}{m}} + \frac{1}{k^{\frac{mn+2}{m}}} c_n^2 x^{n-2} (-b)^2 (x-b)^{\frac{2}{m}} + ... + \frac{1}{k^{\frac{mn+2}{m}}} c_n^{n-1} x (-b)^{n-1} (x-b)^{\frac{2}{m}} + \frac{1}{k^{\frac{mn+2}{m}}} c_n^n (-b)^n (x-b)^{\frac{2}{m}}.$$

Part 3: $f(x) = kx^{\frac{m}{mn+3}} + b; a \in R; b \in R; a \neq 0; b \neq 0; m = 2, 3, 4, ..., \infty; n = 1, 2, 3, ..., \infty$.

Its inverse function is:

$$f^{-1}(x) = \frac{1}{k^{\frac{mn+3}{m}}} c_n^0 x^n (x-b)^{\frac{3}{m}} + \frac{1}{k^{\frac{mn+3}{m}}} c_n^1 x^{n-1} (-b)(x-b)^{\frac{3}{m}} + \frac{1}{k^{\frac{mn+3}{m}}} c_n^2 x^{n-2} (-b)^2 (x-b)^{\frac{3}{m}} + ... + \frac{1}{k^{\frac{mn+3}{m}}} c_n^{n-1} x (-b)^{n-1} (x-b)^{\frac{3}{m}} + \frac{1}{k^{\frac{mn+3}{m}}} c_n^n (-b)^n (x-b)^{\frac{3}{m}}.$$

… … … …

Part m: $f(x) = kx^{\frac{m}{mn+m-1}} + b; a \in R; b \in R; a \neq 0; b \neq 0; m = 2, 3, 4, ..., \infty; n = 1, 2, 3, ..., \infty$.

Its inverse function is:

$$f^{-1}(x) = \frac{1}{k^{\frac{mn+m-1}{m}}} c_n^0 x^n (x-b)^{\frac{m-1}{m}} + \frac{1}{k^{\frac{mn+m-1}{m}}} c_n^1 x^{n-1} (-b)(x-b)^{\frac{m-1}{m}} + \frac{1}{k^{\frac{mn+m-1}{m}}} c_n^2 x^{n-2} (-b)^2 (x-b)^{\frac{m-1}{m}} + ... + \frac{1}{k^{\frac{mn+m-1}{m}}} c_n^{n-1} x (-b)^{n-1} (x-b)^{\frac{m-1}{m}} + \frac{1}{k^{\frac{mn+m-1}{m}}} c_n^n (-b)^n (x-b)^{\frac{m-1}{m}}.$$

Part 4:

This is a supplementary part but it is still incomplete.

1: The Multivariate Dinbakish Equations

$$[1]: \quad a_n\left(\frac{x}{y}\right)^{(n)\frac{i}{k}} + a_{n-1}\left(\frac{x}{y}\right)^{(n-1)\frac{i}{k}} + a_{n-2}\left(\frac{x}{y}\right)^{(n-2)\frac{i}{k}} + \ldots + a_2\left(\frac{x}{y}\right)^{(2)\frac{i}{k}} + a_1\left(\frac{x}{y}\right)^{\frac{i}{k}} + a_0 = 0 \; ;$$

$$[2]: \quad a_n\left(\frac{x}{y}\right)^{(n)\left(-\frac{i}{k}\right)} + a_{n-1}\left(\frac{x}{y}\right)^{(n-1)\left(\frac{i}{k}\right)} + a_{n-2}\left(\frac{x}{y}\right)^{(n-2)\left(-\frac{i}{k}\right)} + \ldots + a_2\left(\frac{x}{y}\right)^{(2)\left(-\frac{i}{k}\right)} + a_1\left(\frac{x}{y}\right)^{-\frac{i}{k}} + a_0 = 0 \; ;$$

$$[3]: \quad a_n\left(\frac{x}{y}\right)^{(n)\frac{\sqrt{i}}{k}} + a_{n-1}\left(\frac{x}{y}\right)^{(n-1)\frac{\sqrt{i}}{k}} + a_{n-2}\left(\frac{x}{y}\right)^{(n-2)\frac{\sqrt{i}}{k}} + \ldots + a_2\left(\frac{x}{y}\right)^{(2)\frac{\sqrt{i}}{k}} + a_1\left(\frac{x}{y}\right)^{\frac{\sqrt{i}}{k}} + a_0 = 0 \; ;$$

$$[4]: \quad a_n\left(\frac{x}{y}\right)^{(n)\left(-\frac{\sqrt{i}}{k}\right)} + a_{n-1}\left(\frac{x}{y}\right)^{(n-1)\left(-\frac{\sqrt{i}}{k}\right)} + a_{n-2}\left(\frac{x}{y}\right)^{(n-2)\left(-\frac{\sqrt{i}}{k}\right)} + \ldots + a_2\left(\frac{x}{y}\right)^{(2)\left(-\frac{\sqrt{i}}{k}\right)} + a_1\left(\frac{x}{y}\right)^{-\frac{\sqrt{i}}{k}} + a_0 = 0 \; ;$$

$$[5]: \quad a_n\left(\frac{x}{y}\right)^{(n)\sqrt{i}} + a_{n-1}\left(\frac{x}{y}\right)^{(n-1)\sqrt{i}} + a_{n-2}\left(\frac{x}{y}\right)^{(n-2)\sqrt{i}} + \ldots + a_2\left(\frac{x}{y}\right)^{(2)\sqrt{i}} + a_1\left(\frac{x}{y}\right)^{\sqrt{i}} + a_0 = 0 \; ;$$

$$[6]: \quad a_n\left(\frac{x}{y}\right)^{(n)\left(-\sqrt{i}\right)} + a_{n-1}\left(\frac{x}{y}\right)^{(n-1)\left(-\sqrt{i}\right)} + a_{n-2}\left(\frac{x}{y}\right)^{(n-2)\left(-\sqrt{i}\right)} + \ldots + a_2\left(\frac{x}{y}\right)^{(2)\left(-\sqrt{i}\right)} + a_1\left(\frac{x}{y}\right)^{-\sqrt{i}} + a_0 = 0 \; .$$

$$i = 2,3,4,\ldots,\infty; \; k = 2,3,4,\ldots,\infty; \; n = 1,2,3,\ldots,\infty; \; y = a_1 x + a_0 \; .$$

$$[7]: \quad a_n\left(\frac{x}{yz}\right)^{(n)\frac{i}{k}} + a_{n-1}\left(\frac{x}{yz}\right)^{(n-1)\frac{i}{k}} + a_{n-2}\left(\frac{x}{yz}\right)^{(n-2)\frac{i}{k}} + \ldots + a_2\left(\frac{x}{yz}\right)^{(2)\frac{i}{k}} + a_1\left(\frac{x}{yz}\right)^{\frac{i}{k}} + a_0 = 0 \; ;$$

$$[8]: \quad a_n\left(\frac{x}{yz}\right)^{(n)\left(-\frac{i}{k}\right)} + a_{n-1}\left(\frac{x}{yz}\right)^{(n-1)\left(-\frac{i}{k}\right)} + a_{n-2}\left(\frac{x}{yz}\right)^{(n-2)\left(-\frac{i}{k}\right)} + \ldots + a_2\left(\frac{x}{yz}\right)^{(2)\left(-\frac{i}{k}\right)} + a_1\left(\frac{x}{yz}\right)^{-\frac{i}{k}} + a_0 = 0 \; ;$$

$$[9]: \quad a_n\left(\frac{x}{yz}\right)^{(n)\frac{\sqrt{i}}{k}} + a_{n-1}\left(\frac{x}{yz}\right)^{(n-1)\frac{\sqrt{i}}{k}} + a_{n-2}\left(\frac{x}{yz}\right)^{(n-2)\frac{\sqrt{i}}{k}} + \ldots + a_2\left(\frac{x}{yz}\right)^{(2)\frac{\sqrt{i}}{k}} + a_1\left(\frac{x}{yz}\right)^{\frac{\sqrt{i}}{k}} + a_0 = 0 \; ;$$

$$[10]: \quad a_n\left(\frac{x}{yz}\right)^{(n)\left(-\frac{\sqrt{i}}{k}\right)} + a_{n-1}\left(\frac{x}{yz}\right)^{(n-1)\left(-\frac{\sqrt{i}}{k}\right)} + a_{n-2}\left(\frac{x}{yz}\right)^{(n-2)\left(-\frac{\sqrt{i}}{k}\right)} + \ldots + a_2\left(\frac{x}{yz}\right)^{(2)\left(-\frac{\sqrt{i}}{k}\right)} + a_1\left(\frac{x}{yz}\right)^{-\frac{\sqrt{i}}{k}} + a_0 = 0 \; ;$$

$$[11]: a_n\left(\frac{x}{yz}\right)^{(n)\sqrt{i}} + a_{n-1}\left(\frac{x}{yz}\right)^{(n-1)\sqrt{i}} + a_{n-2}\left(\frac{x}{yz}\right)^{(n-2)\sqrt{i}} + ... + a_2\left(\frac{x}{yz}\right)^{(2)\sqrt{i}} + a_1\left(\frac{x}{yz}\right)^{\sqrt{i}} + a_0 = 0 \ ;$$

$$[12]: a_n\left(\frac{x}{yz}\right)^{(n)(-\sqrt{i})} + a_{n-1}\left(\frac{x}{yz}\right)^{(n-1)(-\sqrt{i})} + a_{n-2}\left(\frac{x}{yz}\right)^{(n-2)(-\sqrt{i})} + ... + a_2\left(\frac{x}{yz}\right)^{(2)(-\sqrt{i})} + a_1\left(\frac{x}{yz}\right)^{-\sqrt{i}} + a_0 = 0 \ ;$$

$$[13]: a_n\left(\frac{xz}{y}\right)^{(n)\frac{i}{k}} + a_{n-1}\left(\frac{xz}{y}\right)^{(n-1)\frac{i}{k}} + a_{n-2}\left(\frac{xz}{y}\right)^{(n-2)\frac{i}{k}} + ... + a_2\left(\frac{xz}{y}\right)^{(2)\frac{i}{k}} + a_1\left(\frac{xz}{y}\right)^{\frac{i}{k}} + a_0 = 0 \ ;$$

$$[14]: a_n\left(\frac{xz}{y}\right)^{(n)\left(-\frac{i}{k}\right)} + a_{n-1}\left(\frac{xz}{y}\right)^{(n-1)\left(-\frac{i}{k}\right)} + a_{n-2}\left(\frac{xz}{y}\right)^{(n-2)\left(-\frac{i}{k}\right)} + ... + a_2\left(\frac{xz}{y}\right)^{(2)\left(-\frac{i}{k}\right)} + a_1\left(\frac{xz}{y}\right)^{-\frac{i}{k}} + a_0 = 0 \ ;$$

$$[15]: a_n\left(\frac{xz}{y}\right)^{(n)\frac{\sqrt{i}}{k}} + a_{n-1}\left(\frac{xz}{y}\right)^{(n-1)\frac{\sqrt{i}}{k}} + a_{n-2}\left(\frac{xz}{y}\right)^{(n-2)\frac{\sqrt{i}}{k}} + ... + a_2\left(\frac{xz}{y}\right)^{(2)\frac{\sqrt{i}}{k}} + a_1\left(\frac{xz}{y}\right)^{\frac{\sqrt{i}}{k}} + a_0 = 0 \ ;$$

$$[16]: a_n\left(\frac{xz}{y}\right)^{(n)\left(-\frac{\sqrt{i}}{k}\right)} + a_{n-1}\left(\frac{xz}{y}\right)^{(n-1)\left(-\frac{\sqrt{i}}{k}\right)} + a_{n-2}\left(\frac{xz}{y}\right)^{(n-2)\left(-\frac{\sqrt{i}}{k}\right)} + ... + a_2\left(\frac{xz}{y}\right)^{(2)\left(-\frac{\sqrt{i}}{k}\right)} + a_1\left(\frac{xz}{y}\right)^{-\frac{\sqrt{i}}{k}} + a_0 = 0 \ ;$$

$$[17]: a_n\left(\frac{xz}{y}\right)^{(n)\sqrt{i}} + a_{n-1}\left(\frac{xz}{y}\right)^{(n-1)\sqrt{i}} + a_{n-2}\left(\frac{xz}{y}\right)^{(n-2)\sqrt{i}} + ... + a_2\left(\frac{xz}{y}\right)^{(2)\sqrt{i}} + a_1\left(\frac{xz}{y}\right)^{\sqrt{i}} + a_0 = 0 \ ;$$

$$[18]: a_n\left(\frac{xz}{y}\right)^{(n)(-\sqrt{i})} + a_{n-1}\left(\frac{xz}{y}\right)^{(n-1)(-\sqrt{i})} + a_{n-2}\left(\frac{xz}{y}\right)^{(n-2)(-\sqrt{i})} + ... + a_2\left(\frac{xz}{y}\right)^{(2)(-\sqrt{i})} + a_1\left(\frac{xz}{y}\right)^{-\sqrt{i}} + a_0 = 0 \ ;$$

$$i = 2,3,4,...,\infty; k = 2,3,4,...,\infty; n = 1,2,3,...,\infty; y = a_1 x + a_0; z = a_1 x - a_0 \ .$$

2: The Composite-Multivariate Dinbakish Equations

$$[1]: a_{2n}\left(\frac{x}{y}\right)^{(n-1)\frac{i}{k}+\sqrt{j}} + a_{2n-1}\left(\frac{x}{y}\right)^{(n-1)\frac{i}{k}} + a_{2n-2}\left(\frac{x}{y}\right)^{(n-2)\frac{i}{k}+\sqrt{j}} + a_{2n-3}\left(\frac{x}{y}\right)^{(n-2)\frac{i}{k}} + ... + a_3\left(\frac{x}{y}\right)^{\frac{i}{k}} + a_2\left(\frac{x}{y}\right)^{\sqrt{j}} + a_1 = 0 \ ;$$

$$[2]: a_{2n}\left(\frac{x}{y}\right)^{(n-1)\frac{i}{k}-\sqrt{j}} + a_{2n-1}\left(\frac{x}{y}\right)^{(n-1)\frac{i}{k}} + a_{2n-2}\left(\frac{x}{y}\right)^{(n-2)\frac{i}{k}-\sqrt{j}} + a_{2n-3}\left(\frac{x}{y}\right)^{(n-2)\frac{i}{k}} + ... + a_3\left(\frac{x}{y}\right)^{\frac{i}{k}} + a_2\left(\frac{x}{y}\right)^{-\sqrt{j}} + a_1 = 0 \ ;$$

154

$$[3]: \quad a_{2n}\left(\frac{x}{y}\right)^{(n-1)\left(-\frac{i}{k}\right)+\sqrt{j}}+a_{2n-1}\left(\frac{x}{y}\right)^{(n-1)\left(-\frac{i}{k}\right)}+a_{2n-2}\left(\frac{x}{y}\right)^{(n-2)\left(-\frac{i}{k}\right)+\sqrt{j}}+a_{2n-3}\left(\frac{x}{y}\right)^{(n-2)\left(-\frac{i}{k}\right)}+...+a_{3}\left(\frac{x}{y}\right)^{-\frac{i}{k}}+a_{2}\left(\frac{x}{y}\right)^{\sqrt{j}}+a_{1}=0 \ ;$$

$$[4]: \quad a_{2n}\left(\frac{x}{y}\right)^{(n-1)\left(-\frac{i}{k}\right)-\sqrt{j}}+a_{2n-1}\left(\frac{x}{y}\right)^{(n-1)\left(-\frac{i}{k}\right)}+a_{2n-2}\left(\frac{x}{y}\right)^{(n-2)\left(-\frac{i}{k}\right)-\sqrt{j}}+a_{2n-3}\left(\frac{x}{y}\right)^{(n-2)\left(-\frac{i}{k}\right)}+...+a_{3}\left(\frac{x}{y}\right)^{-\frac{i}{k}}+a_{2}\left(\frac{x}{y}\right)^{-\sqrt{j}}+a_{1}=0 \ ;$$

$$[5]: \quad a_{2n}\left(\frac{x}{y}\right)^{(n-1)\frac{i}{k}+\frac{\sqrt{j}}{m}}+a_{2n-1}\left(\frac{x}{y}\right)^{(n-1)\frac{i}{k}}+a_{2n-2}\left(\frac{x}{y}\right)^{(n-2)\frac{i}{k}+\frac{\sqrt{j}}{m}}+a_{2n-3}\left(\frac{x}{y}\right)^{(n-2)\frac{i}{k}}+...+a_{3}\left(\frac{x}{y}\right)^{\frac{i}{k}}+a_{2}\left(\frac{x}{y}\right)^{\frac{\sqrt{j}}{m}}+a_{1}=0 \ ;$$

$$[6]: \quad a_{2n}\left(\frac{x}{y}\right)^{(n-1)\frac{i}{k}-\frac{\sqrt{j}}{m}}+a_{2n-1}\left(\frac{x}{y}\right)^{(n-1)\frac{i}{k}}+a_{2n-2}\left(\frac{x}{y}\right)^{(n-2)\frac{i}{k}-\frac{\sqrt{j}}{m}}+a_{2n-3}\left(\frac{x}{y}\right)^{(n-2)\frac{i}{k}}+...+a_{3}\left(\frac{x}{y}\right)^{\frac{i}{k}}+a_{2}\left(\frac{x}{y}\right)^{-\frac{\sqrt{j}}{m}}+a_{1}=0 \ ;$$

$$[7]: \quad a_{2n}\left(\frac{x}{y}\right)^{(n-1)\left(-\frac{i}{k}\right)+\frac{\sqrt{j}}{m}}+a_{2n-1}\left(\frac{x}{y}\right)^{(n-1)\left(-\frac{i}{k}\right)}+a_{2n-2}\left(\frac{x}{y}\right)^{(n-2)\left(-\frac{i}{k}\right)+\frac{\sqrt{j}}{m}}+a_{2n-3}\left(\frac{x}{y}\right)^{(n-2)\left(-\frac{i}{k}\right)}+...+a_{3}\left(\frac{x}{y}\right)^{-\frac{i}{k}}+a_{2}\left(\frac{x}{y}\right)^{\frac{\sqrt{j}}{m}}+a_{1}=0 \ ;$$

$$[8]: \quad a_{2n}\left(\frac{x}{y}\right)^{(n-1)\left(-\frac{i}{k}\right)-\frac{\sqrt{j}}{m}}+a_{2n-1}\left(\frac{x}{y}\right)^{(n-1)\left(-\frac{i}{k}\right)}+a_{2n-2}\left(\frac{x}{y}\right)^{(n-2)\left(-\frac{i}{k}\right)-\frac{\sqrt{j}}{m}}+a_{2n-3}\left(\frac{x}{y}\right)^{(n-2)\left(-\frac{i}{k}\right)}+...+a_{3}\left(\frac{x}{y}\right)^{-\frac{i}{k}}+a_{2}\left(\frac{x}{y}\right)^{-\frac{\sqrt{j}}{m}}+a_{1}=0 \ ;$$

$$[9]: \quad a_{2n}\left(\frac{x}{y}\right)^{(n-1)\sqrt{i}+\sqrt{j}}+a_{2n-1}\left(\frac{x}{y}\right)^{(n-1)\sqrt{i}}+a_{2n-2}\left(\frac{x}{y}\right)^{(n-2)\sqrt{i}+\sqrt{j}}+a_{2n-3}\left(\frac{x}{y}\right)^{(n-2)\sqrt{i}}+...+a_{3}\left(\frac{x}{y}\right)^{\sqrt{i}}+a_{2}\left(\frac{x}{y}\right)^{\sqrt{j}}+a_{1}=0 \ ;$$

$$[10]: \quad a_{2n}\left(\frac{x}{y}\right)^{(n-1)\sqrt{i}-\sqrt{j}}+a_{2n-1}\left(\frac{x}{y}\right)^{(n-1)\sqrt{i}}+a_{2n-2}\left(\frac{x}{y}\right)^{(n-2)\sqrt{i}-\sqrt{j}}+a_{2n-3}\left(\frac{x}{y}\right)^{(n-2)\sqrt{i}}+...+a_{3}\left(\frac{x}{y}\right)^{\sqrt{i}}+a_{2}\left(\frac{x}{y}\right)^{-\sqrt{j}}+a_{1}=0 \ ;$$

$$[11]: \quad a_{2n}\left(\frac{x}{y}\right)^{(n-1)\left(-\sqrt{i}\right)+\sqrt{j}}+a_{2n-1}\left(\frac{x}{y}\right)^{(n-1)\left(-\sqrt{i}\right)}+a_{2n-2}\left(\frac{x}{y}\right)^{(n-2)\left(-\sqrt{i}\right)+\sqrt{j}}+a_{2n-3}\left(\frac{x}{y}\right)^{(n-2)\left(-\sqrt{i}\right)}+...+a_{3}\left(\frac{x}{y}\right)^{-\sqrt{i}}+a_{2}\left(\frac{x}{y}\right)^{\sqrt{j}}+a_{1}=0 \ ;$$

$$[12]: \quad a_{2n}\left(\frac{x}{y}\right)^{(n-1)\left(-\sqrt{i}\right)-\sqrt{j}}+a_{2n-1}\left(\frac{x}{y}\right)^{(n-1)\left(-\sqrt{i}\right)}+a_{2n-2}\left(\frac{x}{y}\right)^{(n-2)\left(-\sqrt{i}\right)-\sqrt{j}}+a_{2n-3}\left(\frac{x}{y}\right)^{(n-2)\left(-\sqrt{i}\right)}+...+a_{3}\left(\frac{x}{y}\right)^{-\sqrt{i}}+a_{2}\left(\frac{x}{y}\right)^{-\sqrt{j}}+a_{1}=0 \ ;$$

$$[13]: \quad a_{2n}\left(\frac{x}{y}\right)^{(n-1)\sqrt{j}+\frac{i}{k}}+a_{2n-1}\left(\frac{x}{y}\right)^{(n-1)\sqrt{j}}+a_{2n-2}\left(\frac{x}{y}\right)^{(n-2)\sqrt{j}+\frac{i}{k}}+a_{2n-3}\left(\frac{x}{y}\right)^{(n-2)\sqrt{j}}+...+a_{3}\left(\frac{x}{y}\right)^{\sqrt{j}}+a_{2}\left(\frac{x}{y}\right)^{\frac{i}{k}}+a_{1}=0 \ ;$$

$$[14]: \quad a_{2n}\left(\frac{x}{y}\right)^{(n-1)\sqrt{j}-\frac{i}{k}}+a_{2n-1}\left(\frac{x}{y}\right)^{(n-1)\sqrt{j}}+a_{2n-2}\left(\frac{x}{y}\right)^{(n-2)\sqrt{j}-\frac{i}{k}}+a_{2n-3}\left(\frac{x}{y}\right)^{(n-2)\sqrt{j}}+...+a_{3}\left(\frac{x}{y}\right)^{\sqrt{j}}+a_{2}\left(\frac{x}{y}\right)^{-\frac{i}{k}}+a_{1}=0 \ ;$$

$[15]$: $a_{2n}\left(\dfrac{x}{y}\right)^{(n-1)\left(-\sqrt{j}\right)-\frac{i}{k}} + a_{2n-1}\left(\dfrac{x}{y}\right)^{(n-1)\left(-\sqrt{j}\right)} + a_{2n-2}\left(\dfrac{x}{y}\right)^{(n-2)\left(-\sqrt{j}\right)-\frac{i}{k}} + a_{2n-3}\left(\dfrac{x}{y}\right)^{(n-2)\left(-\sqrt{j}\right)} + \ldots + a_3\left(\dfrac{x}{y}\right)^{-\sqrt{j}} + a_2\left(\dfrac{x}{y}\right)^{-\frac{i}{k}} + a_1 = 0$;

$[16]$: $a_{2n}\left(\dfrac{x}{y}\right)^{(n-1)\left(-\sqrt{j}\right)+\frac{i}{k}} + a_{2n-1}\left(\dfrac{x}{y}\right)^{(n-1)\left(-\sqrt{j}\right)} + a_{2n-2}\left(\dfrac{x}{y}\right)^{(n-2)\left(-\sqrt{j}\right)+\frac{i}{k}} + a_{2n-3}\left(\dfrac{x}{y}\right)^{(n-2)\left(-\sqrt{j}\right)} + \ldots + a_3\left(\dfrac{x}{y}\right)^{-\sqrt{j}} + a_2\left(\dfrac{x}{y}\right)^{\frac{i}{k}} + a_1 = 0$;

$[17]$: $a_{2n}\left(\dfrac{x}{y}\right)^{(n-1)\sqrt{j}+\frac{\sqrt{i}}{k}} + a_{2n-1}\left(\dfrac{x}{y}\right)^{(n-1)\sqrt{j}} + a_{2n-2}\left(\dfrac{x}{y}\right)^{(n-2)\sqrt{j}+\frac{\sqrt{i}}{k}} + a_{2n-3}\left(\dfrac{x}{y}\right)^{(n-2)\sqrt{j}} + \ldots + a_3\left(\dfrac{x}{y}\right)^{\sqrt{j}} + a_2\left(\dfrac{x}{y}\right)^{\frac{\sqrt{i}}{k}} + a_1 = 0$;

$[18]$: $a_{2n}\left(\dfrac{x}{y}\right)^{(n-1)\sqrt{j}-\frac{\sqrt{i}}{k}} + a_{2n-1}\left(\dfrac{x}{y}\right)^{(n-1)\sqrt{j}} + a_{2n-2}\left(\dfrac{x}{y}\right)^{(n-2)\sqrt{j}-\frac{\sqrt{i}}{k}} + a_{2n-3}\left(\dfrac{x}{y}\right)^{(n-2)\sqrt{j}} + \ldots + a_3\left(\dfrac{x}{y}\right)^{\sqrt{j}} + a_2\left(\dfrac{x}{y}\right)^{-\frac{\sqrt{i}}{k}} + a_1 = 0$;

$[19]$: $a_{2n}\left(\dfrac{x}{y}\right)^{(n-1)\left(-\sqrt{j}\right)+\frac{\sqrt{i}}{k}} + a_{2n-1}\left(\dfrac{x}{y}\right)^{(n-1)\left(-\sqrt{j}\right)} + a_{2n-2}\left(\dfrac{x}{y}\right)^{(n-2)\left(-\sqrt{j}\right)+\frac{\sqrt{i}}{k}} + a_{2n-3}\left(\dfrac{x}{y}\right)^{(n-2)\left(-\sqrt{j}\right)} + \ldots + a_3\left(\dfrac{x}{y}\right)^{-\sqrt{j}} + a_2\left(\dfrac{x}{y}\right)^{\frac{\sqrt{i}}{k}} + a_1 = 0$;

$[20]$: $a_{2n}\left(\dfrac{x}{y}\right)^{(n-1)\left(-\sqrt{j}\right)-\frac{\sqrt{i}}{k}} + a_{2n-1}\left(\dfrac{x}{y}\right)^{(n-1)\left(-\sqrt{j}\right)} + a_{2n-2}\left(\dfrac{x}{y}\right)^{(n-2)\left(-\sqrt{j}\right)-\frac{\sqrt{i}}{k}} + a_{2n-3}\left(\dfrac{x}{y}\right)^{(n-2)\left(-\sqrt{j}\right)} + \ldots + a_3\left(\dfrac{x}{y}\right)^{-\sqrt{j}} + a_2\left(\dfrac{x}{y}\right)^{-\frac{\sqrt{i}}{k}} + a_1 = 0$.

$$i = 2,3,4,\ldots,\infty; \; j = 2,3,4,\ldots,\infty; \; k = 2,3,4,\ldots,\infty; \; m = 2,3,4,\ldots,\infty; \; n = 1,2,3,\ldots,\infty; \; y = a_2 x + a_1 .$$

3: The Multivariate Dinbakish-Funn Equations

$[1]$: $a_n\left(xy\right)^n + a_{n-1}\left(xy\right)^{n-1} + a_{n-2}\left(xy\right)^{n-2} + \ldots + a_2\left(xy\right)^2 + a_1\left(xy\right) + a_0 = 0$;

$[2]$: $a_n\left(xyz\right)^n + a_{n-1}\left(xyz\right)^{n-1} + a_{n-2}\left(xyz\right)^{n-2} + \ldots + a_2\left(xyz\right)^2 + a_1\left(xyz\right) + a_0 = 0$;

$[3]$: $a_n\left(\dfrac{x}{y}\right)^n + a_{n-1}\left(\dfrac{x}{y}\right)^{n-1} + a_{n-2}\left(\dfrac{x}{y}\right)^{n-2} + \ldots + a_2\left(\dfrac{x}{y}\right)^2 + a_1\left(\dfrac{x}{y}\right) + a_0 = 0$;

$[4]$: $a_n\left(\dfrac{x}{yz}\right)^n + a_{n-1}\left(\dfrac{x}{yz}\right)^{n-1} + a_{n-2}\left(\dfrac{x}{yz}\right)^{n-2} + \ldots + a_2\left(\dfrac{x}{yz}\right)^2 + a_1\left(\dfrac{x}{yz}\right) + a_0 = 0$;

$[5]$: $a_n\left(\dfrac{xz}{y}\right)^n + a_{n-1}\left(\dfrac{xz}{y}\right)^{n-1} + a_{n-2}\left(\dfrac{xz}{y}\right)^{n-2} + \ldots + a_2\left(\dfrac{xz}{y}\right)^2 + a_1\left(\dfrac{xz}{y}\right) + a_0 = 0$;

$$n = 1, 2, 3, ..., \infty; \quad y = a_1 x^{\frac{1}{2}} + a_0; \quad z = a_1 x^{\frac{1}{2}} - a_0 \; .$$

4: The Composite-Multivariate Dinbakish-Funn Equations

$[1]$: $a_{2n}(xy)^{(n-1)j+\frac{i}{k}} + a_{2n-1}(xy)^{(n-1)j} + a_{2n-2}(xy)^{(n-2)j+\frac{i}{k}} + a_{2n-3}(xy)^{(n-2)j} + ... + a_3(xy)^{j} + a_2(xy)^{\frac{i}{k}} + a_1 = 0$;

$[2]$: $a_{2n}(xy)^{(n-1)j-\frac{i}{k}} + a_{2n-1}(xy)^{(n-1)j} + a_{2n-2}(xy)^{(n-2)j-\frac{i}{k}} + a_{2n-3}(xy)^{(n-2)j} + ... + a_3(xy)^{j} + a_2(xy)^{-\frac{i}{k}} + a_1 = 0$;

$[3]$: $a_{2n}(xy)^{(n-1)(-j)+\frac{i}{k}} + a_{2n-1}(xy)^{(n-1)(-j)} + a_{2n-2}(xy)^{(n-2)(-j)+\frac{i}{k}} + a_{2n-3}(xy)^{(n-2)(-j)} + ... + a_3(xy)^{-j} + a_2(xy)^{\frac{i}{k}} + a_1 = 0$;

$[4]$: $a_{2n}(xy)^{(n-1)(-j)-\frac{i}{k}} + a_{2n-1}(xy)^{(n-1)(-j)} + a_{2n-2}(xy)^{(n-2)(-j)-\frac{i}{k}} + a_{2n-3}(xy)^{(n-2)(-j)} + ... + a_3(xy)^{-j} + a_2(xy)^{-\frac{i}{k}} + a_1 = 0$;

$[5]$: $a_{2n}(xy)^{(n-1)j+\frac{\sqrt{i}}{k}} + a_{2n-1}(xy)^{(n-1)j} + a_{2n-2}(xy)^{(n-2)j+\frac{\sqrt{i}}{k}} + a_{2n-3}(xy)^{(n-2)j} + ... + a_3(xy)^{j} + a_2(xy)^{\frac{\sqrt{i}}{k}} + a_1 = 0$;

$[6]$: $a_{2n}(xy)^{(n-1)j-\frac{\sqrt{i}}{k}} + a_{2n-1}(xy)^{(n-1)j} + a_{2n-2}(xy)^{(n-2)j-\frac{\sqrt{i}}{k}} + a_{2n-3}(xy)^{(n-2)j} + ... + a_3(xy)^{j} + a_2(xy)^{-\frac{\sqrt{i}}{k}} + a_1 = 0$;

$[7]$: $a_{2n}(xy)^{(n-1)(-j)+\frac{\sqrt{i}}{k}} + a_{2n-1}(xy)^{(n-1)(-j)} + a_{2n-2}(xy)^{(n-2)(-j)+\frac{\sqrt{i}}{k}} + a_{2n-3}(xy)^{(n-2)(-j)} + ... + a_3(xy)^{-j} + a_2(xy)^{\frac{\sqrt{i}}{k}} + a_1 = 0$;

$[8]$: $a_{2n}(xy)^{(n-1)(-j)-\frac{\sqrt{i}}{k}} + a_{2n-1}(xy)^{(n-1)(-j)} + a_{2n-2}(xy)^{(n-2)(-j)-\frac{\sqrt{i}}{k}} + a_{2n-3}(xy)^{(n-2)(-j)} + ... + a_3(xy)^{-j} + a_2(xy)^{-\frac{\sqrt{i}}{k}} + a_1 = 0$;

$[9]$: $a_{2n}(xy)^{(n-1)j+\sqrt{i}} + a_{2n-1}(xy)^{(n-1)j} + a_{2n-2}(xy)^{(n-2)j+\sqrt{i}} + a_{2n-3}(xy)^{(n-2)j} + ... + a_3(xy)^{j} + a_2(xy)^{\sqrt{i}} + a_1 = 0$;

$[10]$: $a_{2n}(xy)^{(n-1)j-\sqrt{i}} + a_{2n-1}(xy)^{(n-1)j} + a_{2n-2}(xy)^{(n-2)j-\sqrt{i}} + a_{2n-3}(xy)^{(n-2)j} + ... + a_3(xy)^{j} + a_2(xy)^{-\sqrt{i}} + a_1 = 0$;

$[11]$: $a_{2n}(xy)^{(n-1)(-j)+\sqrt{i}} + a_{2n-1}(xy)^{(n-1)(-j)} + a_{2n-2}(xy)^{(n-2)(-j)+\sqrt{i}} + a_{2n-3}(xy)^{(n-2)(-j)} + ... + a_3(xy)^{-j} + a_2(xy)^{\sqrt{i}} + a_1 = 0$;

$[12]$: $a_{2n}(xy)^{(n-1)(-j)-\sqrt{i}} + a_{2n-1}(xy)^{(n-1)(-j)} + a_{2n-2}(xy)^{(n-2)(-j)-\sqrt{i}} + a_{2n-3}(xy)^{(n-2)(-j)} + ... + a_3(xy)^{-j} + a_2(xy)^{-\sqrt{i}} + a_1 = 0$;

$[13]$: $a_{2n}(xy)^{(n-1)\frac{i}{k}+j} + a_{2n-1}(xy)^{(n-1)\frac{i}{k}} + a_{2n-2}(xy)^{(n-2)\frac{i}{k}+j} + a_{2n-3}(xy)^{(n-2)\frac{i}{k}} + ... + a_3(xy)^{\frac{i}{k}} + a_2(xy)^{j} + a_1 = 0$;

$[14]$: $a_{2n}(xy)^{(n-1)\frac{i}{k}-j} + a_{2n-1}(xy)^{(n-1)\frac{i}{k}} + a_{2n-2}(xy)^{(n-2)\frac{i}{k}-j} + a_{2n-3}(xy)^{(n-2)\frac{i}{k}} + ... + a_3(xy)^{\frac{i}{k}} + a_2(xy)^{-j} + a_1 = 0$;

$[15]$: $a_{2n}(xy)^{(n-1)\left(-\frac{i}{k}\right)+j} + a_{2n-1}(xy)^{(n-1)\left(-\frac{i}{k}\right)} + a_{2n-2}(xy)^{(n-2)\left(-\frac{i}{k}\right)+j} + a_{2n-3}(xy)^{(n-2)\left(-\frac{i}{k}\right)} + ... + a_3(xy)^{-\frac{i}{k}} + a_2(xy)^{j} + a_1 = 0$;

$[16]$: $a_{2n}(xy)^{(n-1)\left(-\frac{i}{k}\right)-j}+a_{2n-1}(xy)^{(n-1)\left(-\frac{i}{k}\right)}+a_{2n-2}(xy)^{(n-2)\left(-\frac{i}{k}\right)-j}+a_{2n-3}(xy)^{(n-2)\left(-\frac{i}{k}\right)}+...+a_3(xy)^{-\frac{i}{k}}+a_2(xy)^{-j}+a_1=0$;

$[17]$: $a_{2n}(xy)^{(n-1)\frac{\sqrt{i}}{k}+j}+a_{2n-1}(xy)^{(n-1)\frac{\sqrt{i}}{k}}+a_{2n-2}(xy)^{(n-2)\frac{\sqrt{i}}{k}+j}+a_{2n-3}(xy)^{(n-2)\frac{\sqrt{i}}{k}}+...+a_3(xy)^{\frac{\sqrt{i}}{k}}+a_2(xy)^{j}+a_1=0$;

$[18]$: $a_{2n}(xy)^{(n-1)\frac{\sqrt{i}}{k}-j}+a_{2n-1}(xy)^{(n-1)\frac{\sqrt{i}}{k}}+a_{2n-2}(xy)^{(n-2)\frac{\sqrt{i}}{k}-j}+a_{2n-3}(xy)^{(n-2)\frac{\sqrt{i}}{k}}+...+a_3(xy)^{\frac{\sqrt{i}}{k}}+a_2(xy)^{-j}+a_1=0$;

$[19]$: $a_{2n}(xy)^{(n-1)\left(-\frac{\sqrt{i}}{k}\right)+j}+a_{2n-1}(xy)^{(n-1)\left(-\frac{\sqrt{i}}{k}\right)}+a_{2n-2}(xy)^{(n-2)\left(-\frac{\sqrt{i}}{k}\right)+j}+a_{2n-3}(xy)^{(n-2)\left(-\frac{\sqrt{i}}{k}\right)}+...+a_3(xy)^{-\frac{\sqrt{i}}{k}}+a_2(xy)^{j}+a_1=0$;

$[20]$: $a_{2n}(xy)^{(n-1)\left(-\frac{\sqrt{i}}{k}\right)-j}+a_{2n-1}(xy)^{(n-1)\left(-\frac{\sqrt{i}}{k}\right)}+a_{2n-2}(xy)^{(n-2)\left(-\frac{\sqrt{i}}{k}\right)-j}+a_{2n-3}(xy)^{(n-2)\left(-\frac{\sqrt{i}}{k}\right)}+...+a_3(xy)^{-\frac{\sqrt{i}}{k}}+a_2(xy)^{-j}+a_1=0$;

$[21]$: $a_{2n}(xy)^{(n-1)\sqrt{i}+j}+a_{2n-1}(xy)^{(n-1)\sqrt{i}}+a_{2n-2}(xy)^{(n-2)\sqrt{i}+j}+a_{2n-3}(xy)^{(n-2)\sqrt{i}}+...+a_3(xy)^{\sqrt{i}}+a_2(xy)^{j}+a_1=0$;

$[22]$: $a_{2n}(xy)^{(n-1)\sqrt{i}-j}+a_{2n-1}(xy)^{(n-1)\sqrt{i}}+a_{2n-2}(xy)^{(n-2)\sqrt{i}-j}+a_{2n-3}(xy)^{(n-2)\sqrt{i}}+...+a_3(xy)^{\sqrt{i}}+a_2(xy)^{-j}+a_1=0$;

$[23]$: $a_{2n}(xy)^{(n-1)\left(-\sqrt{i}\right)+j}+a_{2n-1}(xy)^{(n-1)\left(-\sqrt{i}\right)}+a_{2n-2}(xy)^{(n-2)\left(-\sqrt{i}\right)+j}+a_{2n-3}(xy)^{(n-2)\left(-\sqrt{i}\right)}+...+a_3(xy)^{-\sqrt{i}}+a_2(xy)^{j}+a_1=0$;

$[24]$: $a_{2n}(xy)^{(n-1)\left(-\sqrt{i}\right)-j}+a_{2n-1}(xy)^{(n-1)\left(-\sqrt{i}\right)}+a_{2n-2}(xy)^{(n-2)\left(-\sqrt{i}\right)-j}+a_{2n-3}(xy)^{(n-2)\left(-\sqrt{i}\right)}+...+a_3(xy)^{-\sqrt{i}}+a_2(xy)^{-j}+a_1=0$.

Ha, do you all remember this equation $x^5+3x^4-3x^3-9x^2+2x+6=0$?

Its other roots are: 1) $x_1=\sqrt{2}$; 2) $x_2=-\sqrt{2}$; 3) $x_3=-3$

www.ingramcontent.com/pod-product-compliance
Lightning Source LLC
Chambersburg PA
CBHW082012230526
45468CB00022B/1982